ARCHBISHOP RIORDAN HIGH SCHOOL
6532
Laying waste
P9-CCW-650

LAYING WASTE

MICHAEL H. BROWN

PANTHEON BOOKS · NEW YORK

THE POISONING OF AMERICA BY TOXIC CHEMICALS

 Published in the United States by Pantheon Books, a division of Random House, Inc., New York, and simultaneously in Canada by Random House of Canada Limited, Toronto.

LIBRARY OF CONGRESS CATALOGING IN PUBLICATION DATA
Brown, Michael Harold, 1952–
Laying waste.

Includes index.
1. Hazardous wastes—United States. 2. Hazardous wastes—New York (State)—Niagara Falls (City)
I. Title.
TD811.5.B76 614.7′6′0976 79-3311
ISBN 0-394-50808-4

Publisher's Note: This work contains reports of civil and criminal proceedings brought against corporations and individuals for alleged violations of state and federal antipollution statutes.

While every effort has been made to report accurately the status of those prosecutions, changes may have occurred between the time this work was completed and its publication.

Grateful acknowledgment is made for permission to reprint material from the following:

The Texas Monthly, Austin, Texas

The White Lake Observer, Montague, Michigan

Letter to Congressman George E. Brown from Dr. Jerrold L. Wheaton, Director of Health, Riverside county, California

Letter from Mr. William Mattox, Riverside, California

Designed by Susan Mitchell

Manufactured in the United States of America
FIRST EDITION

FOR MY PARENTS

CONTENTS

ACKNOWLEDGMENTS

There are many people to whom I owe a great deal of gratitude. The book could have developed only with the unusual insights, patience, and moral convictions of Phil Pochoda, my editor at Pantheon, Virginia Barber and Mary Evans, my agents, and Holcombe Noble, an editor at the *New York Times Magazine*. The copyediting of Jeanne Morton was invaluable. Of further assistance was Helen English, an attorney at Random House. I was first put on to this issue by Joan Gipp, a council member in the town of Lewiston, New York, who has probably worked longer and harder than anyone else in the nation in exposing toxic waste problems. I am indebted to Karen Schroeder of the Love Canal, whose protests about conditions there were the first and whose keen memory was essential to my early reporting. I would also like to thank Congressman John La Falce and his assistant, Bonnie Casper; Dr. David Axelrod and Marvin Nailor of the New York State Health Department; Lois Gibbs, Marie Pozniak, and Grace McCoulf of the Love Canal Homeowners Association; chemists Edward Kleppinger, Steve Odojewski, and Randy Rakocynski; Dr. James Truchan; Hugh Kaufman; Carol Jean Kruger; Aileen Voorhees; James Clark; Patricia Pino; Dr. Stephen Kim; Paul Chenard; David Ewell; the Sterling family; Robert Manning and C. Michael Curtis of the *Atlantic Monthly*; and many others whose thoughts and experiences were so generously lent. Extremely helpful for background factual content were Rachel Carson's *Silent Spring*, Dr. Samuel Epstein's *The Politics of Cancer*, Dr. George Waldbott's *Health Effects of Environmental Pollutants*, and the scientific papers of Dr. Pietro Capurro.

INTRODUCTION

When I returned to my home city of Niagara Falls in 1975 after a five-year absence, there was no hint of the horrors I would soon witness. On the surface it still appeared the famous tourist city, and after fifty long years in the grip of decline and renovation efforts, it now was trying to get back on the track of success. Local politicians were touting the marvelous new buildings they said would soon grace the barren downtown, once again attracting hordes of sightseers. A large new convention center had opened in the middle of the city. While the chemical factories still filled the air with repugnant odors, the smokestacks seemed to be releasing less black stuff than they once had, and polluted Lake Erie was taking a turn for the better.

At the beginning of 1977, having just completed my first book, I joined the local newspaper, the *Niagara Gazette*, as a reporter. Soon I discovered that a new chemical facility was operating in a suburb of Niagara Falls. Never before had I known of that kind of plant, for its sole duty was to receive waste materials from other factories and dispose of them. It was doing this in huge ground excavations hard by Lake Ontario. While at first I assumed that the company represented the modern means of disposal, a progressive alternative to pouring dregs directly into sewers, further research and reflection convinced me that its methods merely delayed the negative impacts of waste chemicals. Eventually those buried materials would find their way through the soil and into the lake.

Soon there developed a heated debate over the existence of such a disposal firm in Niagara County, and at one public hearing in 1977 a woman of about twenty got up before the microphone to argue against allowing the waste plant to remain. I watched perplexedly as she began to cry upon mentioning a similar chemical dumpsite in the city of Niagara Falls which, she said, was damaging her neighborhood. She referred to the dump as the "Love Canal." From several news clippings in the newspaper's morgue written by David Pollak, I learned that the "Love Canal" was an old chemical depository located in a residential part of town and once owned by a huge local chemical firm, the Hooker Chemical Company. It was now leaking. The issue, however, appeared to be well in the hands of the authorities, and I decided that the woman's reaction had been based more on emotion than on facts.

My initial perception was wrong. During 1977 and throughout 1978, I began to investigate the Love Canal situation and other instances of ground pollution in the county. The results of that study, which kept me on the newspaper much longer than I had planned, are presented in the first section of this book. Contrary to my first impressions of a "new Niagara," the city was being devastated by factory wastes buried under the ground. A good number of those who, like myself, had been born in Niagara, and who had faith in its future, were under assault from buried wastes. I saw homes where dogs had lost their fur. I saw children with serious birth defects. I saw entire families in inexplicably poor health. When I walked on the Love Canal, I gasped for air as my lungs heaved in fits of wheezing. My eyes burned. There was a sour taste in my mouth. It seemed inconceivable that industry and government could have allowed this to happen, and yet there it was, an exposed cesspool of chemicals threatening not only those who lived nearby but, through its seepage, the viability of the very river my hometown was famous for.

It was not a pleasant task to probe the matter. Day in and day out, at night and on weekends, I tallied endless accounts of health problems in Niagara Falls and watched some people deteriorate before my eyes. Yet those county and city officials to whom we trust our collective well-being repeatedly downplayed the troubles and even subtly discouraged me from pursuing and reporting on them. The city was economically dependent on the chemical trade; health seemed a secondary consideration. In the socioeconomic milieu of Niagara, Hooker was *primus inter pares*. And in my own newsroom, there seemed to be an unwritten law that a reporter did not attack or otherwise fluster the Hooker executives. In such a climate, my stories tended to lack the support warranted by what I discovered. It was not until I contacted the *New York Times Magazine* in June 1978 that I received a sympathetic hearing on what Hooker had done to this place I called home—a city that now seemed polluted to the point of no return. Later, the scandal was fully exposed.

Nor was Niagara alone. Early in 1978, I had begun calling other regions of the country to see if they too had landfill problems. I noticed two major trends: first, that officials, whether on the state or the federal level, knew virtually nothing of or did not care about these problems, and second, that there were indeed other dumpsites leaking their contents in a way that threatened human health.

Since the 1960s, national awareness of the industrial pollution of our air and water had grown enormously; we could see the smoke from chimneys or watch sudsy water collecting near outfall pipes. Yet all the while, other, graver ecological threats were accelerating almost unnoticed. The direct discharge of pollutants had certainly marred our rivers and streams, but the largest quantities of the most hazardous residues—chlorinated pesticides, solvents, and radioactive material—were being placed in containers, stored in old warehouses or on rural land, and eventually dumped into ground pits or caverns throughout the United States, from which they would one day migrate, forgotten by government and unseen by the public. Already land-dumped chemicals were threatening the water supply near Rehoboth, Massachusetts, and discarded chemical drums reportedly had killed thousands of fish in Tennessee's Boone Reservoir. No part of the country was safe from the toxic residues accumulating, unnoticed and untended, at a disastrous rate.

Chemicals have pervaded our lives. We brush our teeth with fluoride compounds, rub on propylene glycol deodorants, clothe ourselves in rayon and nylon or treated cotton and wool, drive cars filled with the products of a liver carcinogen called vinyl chloride, talk on plastic phones, walk on synthetic tiles, live within walls coated with chemical-laden paint. Our food, kept fresh in refrigerators by heat-absorbent refrigerants, contains preservatives and chemical additives. And of course, it has been grown with the aid of chemical fertilizers and insecticides. The detergents, the medicines, the foam rubber, and the floor cleaners—all had their underside of waste, about to make a dramatic public re-entry.

Nature was caught off guard, utterly unprepared for this onslaught of artificial elements. Nowhere in the earth's crust nor in the ocean was there the capability of disassembling these complex new substances strung together by the ingenuity of man. Many of the chemicals, sophisticated hydrocarbons such as DDT, were based on long molecular chains solid enough to withstand degradation by sunlight, dissolution by water, or breakdown by acids, clay minerals, or metal ions. These chemicals, deadly to the human metabolism, found the earth and its waters a congenial way station; and now they remained permanently available to exact a terrible price for human indifference and greed.

I

THE LEGACY OF THE HOOKER CHEMICAL COMPANY

1

LOVE'S NEW MODEL CITY

At each turn, the schizophrenia of Niagara Falls, New York, is starkly evident. A city of unmatched natural beauty, it is also a tired industrial workhorse, beaten often and with a hard hand. In summertime the tourist limousines, heading toward the spectacular falls, move alongside soiled factory tank trucks. All the vehicles traverse a pavement film-coated with oils and soot, past large T-shaped steel constructions that string electrical lines above the densely wooded ravines to the horizon. In the southwest, a rising mist of spray from the downward force of the cataracts contends for prominence in the skyline with the dark plumes of towering smokestacks.

These contradictions stem from the magnificent river—a strait of water, really—that connects Lake Erie to Lake Ontario. Flowing north at a pace of half a million tons a minute, the watercourse widens into a smooth expanse near the city before it breaks into whitecaps and takes its famous 186-foot plunge. Then it cascades through a gorge of overhanging shale and limestone to haystack rapids higher and swifter than anywhere else on the continent. From there, it turns mellow again.

Newlyweds and other tourists have long treated the falls as an obligatory pilgrimage, and they once had come in long lines during the warmer months. At the same time, the plunging river provides cheap electricity for industry, particularly chemical producers, so that a good stretch of its beautiful shoreline is now filled with the spiraled pipes of distilleries. The odors of chlorine and sulfides hang in the air.

A major proportion of those who live in the city of Niagara Falls work in chemical plants, the largest owned by the Hooker Chemical Company. Timothy Schroeder did not. He was a cement technician by trade, dealing with the factories only if they needed a pathway poured or a small foundation set. Tim and his wife, Karen, lived on 99th Street in a ranch-style home with a brick and wood exterior. They had saved all they could to redecorate the inside and to make additions, such as a cement patio covered with an extended roof. One of the Schroeders' most cherished possessions was a fiberglass pool, built into the ground and enclosed by a redwood fence. Though it had taxed their resources, the yard complemented a house that was among the most elegant in a residential zone where most of the homes were small frame buildings, prefabricated and slapped together *en masse*. It was a quiet area, once almost rural in character, and located in the city's extreme southeast corner. The Schroeders had lived in the house only since 1970, but Karen was a lifelong resident of the general neighborhood. Her parents lived three doors down from them, six miles from the row of factories that stood shoulder to shoulder along the Upper Niagara.

Karen Schroeder looked out from a back window one October morning in 1974 and noted with distress that the pool had suddenly risen two feet above the ground. She called Tim to tell him about it. Karen then had no way of knowing that the problem far exceeded a simple property loss—that in fact it was the first sign of a great tragedy.

Accurately enough, Mrs. Schroeder figured that the cause of the uplift was the unusual groundwater flow of the area. Twenty-one years before, an abandoned hydroelectric canal directly behind their house had been backfilled with industrial rubble. The underground breaches created by this disturbance, aided by the marshy nature of the region's surficial layer, had collected large volumes of rainfall, and this water had undermined the backyard. The Schroeders allowed the pool to remain in its precarious position until the next summer and then pulled it from the ground, intending to replace it with a cement one. Immediately, the gaping hole filled with what Karen called "chemical water," rancid liquids of yellow and orchid and blue. These same chemicals, mixed with the groundwater, had flooded the entire yard; they attacked the redwood posts with such a caustic bite that one day the fence simply collapsed. When the groundwater receded in dry weather, it left the gardens and shrubs withered and scorched, as if by a brush fire.

How the chemicals had got there was no mystery: they came from the former canal. Beginning in the late 1930s or the early 1940s, the Hooker Company, whose many processes included the manufacture of pesticides, plasticizers, and caustic soda, had used the canal as a dump for at least 20,000 tons of waste residues—"still-bottoms" in the language of the trade. The chemical garbage was brought to the excavation in 55-gallon metal barrels stacked on a small dump truck and was unloaded into what, up to that time, had been a fishing and swimming hole in the summer and an ice-skating rink during the city's long, hard winter months.

When the hazardous dumping first began, much of the surrounding terrain was meadowlands and orchards, but there was also a small cluster of homes on the immediate periphery, only thirty feet from the ditch. Those who lived there remembered the deep holes being filled with what appeared to be oil and gray mud by laborers who rushed to borrow their garden hoses for a dousing of water if they came in contact with the scalding sludge they were dumping. Children enjoyed playing among the intriguing, unguarded debris. They would pick up chunks of phosphorus and heave them against cement. Upon impact the "fire rocks," as they were called, would brilliantly explode, sending off a trail of white sparks. Fires and explosions erupted spontaneously when the weather was especially hot. Odors similar to those of the industrial districts wafted into the adjacent windows, accompanied by gusts of fly ash. On a humid moonlit night, residents would look toward the canal and see, in the haze above the soil, a greenish luminescence.

Karen's parents had been the first to experience problems with seepage from the canal. In 1959, her mother, Aileen Voorhees, noticed a strange black sludge bleeding through the basement walls. For the next twenty years, she and her husband, Edwin, tried various methods of halting the irritating intrusion, coating the cinder-block walls with sealants and even constructing a gutter along them to intercept the inflow. Nothing could stop a smell like that of a chemical plant from permeating the entire household, and neighborhood calls to the city for help were unavailing. One day, when Edwin punched a hole in the wall to see what was happening, quantities of black liquid poured out. The cinder blocks were full of the stuff.

Although later it was to be determined that they were in imminent danger, the Voorhees treated the problem at first as a mere nuisance. That it involved chemicals, industrial chemicals, was not particularly significant to them. All their life, all of everyone's life in the city, malodorous fumes had been a normal ingredient of the surrounding air.

More ominous than the Voorhees' basement seepage was an event that occurred in the Schroeder family at 11:12 P.M. on November 21, 1968. Karen gave birth to her third child, a seven-pound girl named Sheri. But no sense of elation filled the delivery room, for the baby was born with a heart that beat irregularly and had a hole in it, bone blockages of the nose and partial deafness, deformed external ears, and a cleft palate. By the age of two, it became obvious that the child was mentally retarded. When her teeth came in, there was a double row of them at the bottom. She also developed an enlarged liver.

The Schroeders looked upon these health problems, as well as certain illnesses among their other children, as acts of capricious genes, a vicious quirk of nature. Like Aileen and Edwin Voorhees, they were mainly aware that the chemicals were devaluing their property. The crab-apple tree and evergreens in the back were dead, and even the oak in the front of the house was sick; one year, the leaves fell off on Father's Day.

The canal was dug with much fanfare in the late nineteenth century by a flamboyant entrepreneur named William T. Love. Love arrived in town with a grandiose dream: to construct a carefully planned industrial city with ready access to water power and major markets. The setting for Love's dream was to be a navigable power channel that would extend seven miles from the Upper Niagara near what is now 99th Street to a terrace known as the Niagara Escarpment, where the water would fall 280 feet, circumventing the treacherous falls and at the same time providing cheap power. A city would be constructed near the point where the canal fed back into the river, and it would accommodate 200,000 to 1 million people, he promised. Love's sales speeches were accompanied by advertisements, circulars, and brass bands, with a chorus singing a special ditty to the tune of "Yankee Doodle": "Everybody's come to town,/Those left we all do pity,/For we'll have a jolly time/At Love's new Model City."

So fired by Love's imagination were the state's leaders that they allowed him the rare opportunity of addressing a joint session of the senate and assembly. He was given a free hand to condemn as much property as he liked and to divert whatever amounts of water. But Love's dream quickly became Love's folly, and, financially depleted, he abandoned the project after a mile-long trench, 10 to 40 feet deep and generally 15 yards wide, had been scoured perpendicular to the Niagara River. Eventually the site was acquired by Hooker.

Except for the frivolous history of Mr. Love, and some general information on the chemicals, little was known publicly about the dump in 1977. Few of those who lived in the numerous houses that had sprung up by the site were aware that the large barren field behind them was a burial ground for toxic wastes. That year, while working as a reporter for a local newspaper, the *Niagara Gazette*, I began to inquire regularly about the strange conditions on 99th Street. The Niagara County Health Department and the city both said it was a nuisance condition but no serious danger to the people. The Hooker Company refused to comment on their chemicals, claiming only that they had no records of the burials and that the problem was not their responsibility. In fact, Hooker had deeded the land to the Niagara Falls Board of Education in 1953 for a token $1. At that time the company issued no detailed warnings about the chemicals; a brief paragraph in the quitclaim document disclaimed company liability for any injuries or deaths that might occur at the site. Ralph

Boniello, the board's attorney, said he had never received any phone calls or letters specifically describing the exact nature of the refuse and its potential effects, nor was there, as the company was later to claim, any threat of property condemnation by the board in order to secure the land. "We had no idea what was in there," Boniello said.

Though surely Hooker must have been relieved to rid itself of the contaminated land, when I read its deed I was left with the impression that the wastes there would be a hazard only if physically touched or swallowed. Otherwise, they did not seem to be an overwhelming concern. In reality, the dangers of these wastes far exceeded those of acids or alkalines or inert salts. We now know that the drums dumped in the canal contained a veritable witch's brew of chemistry, compounds of truly remarkable toxicity. There were solvents that attacked the heart and liver, and residues from pesticides so dangerous that their commercial sale had subsequently been restricted or banned outright by the government; some of them are strongly suspected of causing cancer.

Yet Hooker gave no more than a hint of that. When approached by the educational board for the parcel of property it wanted for a new school, B. Klaussen, then Hooker's executive vice-president, replied in a letter to the board:

> Our officers have carefully considered your request. We are very conscious of the need for new elementary schools and realize that the sites must be carefully selected so that they will best serve the area involved. We feel that the board of education has done a fine job in meeting the expanding demand for additional facilities and we are anxious to cooperate in any proper way. We have, therefore, come to the conclusion that since this

location is the most desirable one for this purpose, we will be willing to donate the entire strip between Colvin Boulevard and Frontier Avenue to be used for the erection of a school at a location to be determined. . . .

The school board, apparently unaware of the exact nature of the substances underneath this generously donated property, and woefully incurious, began to build the new school and playground at the canal's midsection. Construction progressed even after the workers struck a drainage trench that gave off a strong chemical odor and then discovered a waste pit nearby. Instead of halting the work, the board simply had the school site moved 80 feet away. Young families began to settle in increasing numbers alongside the dump; many of them had been told that the field was to be a park and recreation area for their children.

If the children found the "playground" interesting, there were times they found it painful as well. When they played on this land that Hooker implied was so well suited for a school, they sneezed and their eyes teared. In the days when dumping was still in progress, they swam at the opposite end of the canal, at times arriving home with hard pimples on their bodies. And Hooker knew that children were playing on its spoils. In 1958, the company was made aware that three children had been burned by exposed residues on the surface of the canal, much of which, according to the residents, had been covered over with nothing more than fly ash and loose dirt. Because it wished to avoid legal repercussions, the company chose not to issue a public warning of the dangers only it could have known were there, nor to have its chemists explain to the people that their homes would have been better placed elsewhere.

The Love Canal was simply unfit to be a container for

hazardous substances, even by the standards of the day, and now, in 1977, the local authorities were belatedly finding that out. Several years of heavy snowfall and rain had filled the sparsely covered channel like a sponge. The contents were overflowing at a frightening rate, seeping readily into the clay, silt, and sandy loam and finding their way through old creekbeds and swales into the neighborhood.

The city of Niagara Falls, I was assured, was planning a remedial drainage program to reduce chemical migration off the site. But it was apparent that no sense of urgency had been attached to the plan, and it was stalled in a ball of red tape. There was hopeless disagreement over who should pay the bill—the city, Hooker, or the board of education—and the engineers seemed confused as to what exactly needed to be done for a problem that had never been confronted elsewhere.

At a meeting in Buffalo during the summer of 1977, I cornered an independent consultant for the city and requested more information on the dump and the proposed remedial action.

"We're not really sure what the final solution should be," he said. "You can't be sure until you know what you're dealing with."

Was there a chance of harm to the people?

He shrugged his shoulders.

How were the potential dangers to be searched out?

"Someone's going to have to dig there and take a good look," he answered. "If they don't, your child or your children's children are going to run into problems."

The same questions were repeated for months, with no answers. Despite the uncertainty of the city's own consultant, the city manager, Donald O'Hara, persisted in his view that the Love Canal, however displeasing to the eyes and nasal passages, was not a crisis but mainly a matter of aesthetics. O'Hara was pleased to remind me that Dr. Francis Clifford, the county health commissioner, supported his opinion. Besides making light of the seepage, O'Hara created an aura of secrecy around information regarding the canal. His concerns appeared to be financial and legal in nature. As manager, O'Hara had pulled the city out from under a staggering debt, and suddenly, with hardly a moment to enjoy a widely publicized budget surplus, his city hall was faced with the prospect of spending an unplanned $400,000 for a remedial program at the dumpsite. And it was feared there would be more expensive work to do later on—and lawsuits.

With the city, the school board, and Hooker unwilling to commit themselves to a remedy, conditions between 97th and 99th streets continued to degenerate until, by early 1978, the land was a quagmire of sludge. Melting snow drained a layer of soot onto the private yards, while the ground on the dump itself had softened to the point of collapse, exposing the crushed tops of barrels. When a city truck attempted to cross the field and dump clay on one especially large hole, it sank up to its axles. Masses of sludge beneath the surface were finding their way out at a quickening rate, forming constant springs of contaminated liquid. So miserable had the Schroeder backyard become that the family gave up trying to fight the inundation. They had brought in an old bulldozer to attempt to cover pools of chemicals that welled up here and there, but now the machine sat still. Their yard, once featured in a local newspaper for its beauty, now had degenerated to the point where it was unfit even to walk upon. Of course, the Schroeders could not leave. No one would think of buying the property. They had a mortgage to pay, and on Tim's salary, could not afford to maintain the house while they moved to a safer setting. They and their four children were stuck.

That the city might be saddled with large costs was not the sole reason for its reluctance to address the issue directly and help the Schroeders and the hundred or so other families whose properties abutted the covered trench. I felt there was also trepidation of a political sort: the fear of distressing Hooker. To an economically depressed area the company provided desperately needed employment—as many as 3,000 blue-collar jobs in the general vicinity at certain periods—and a substantial number of tax dollars. More to the point, perhaps, Hooker was speaking of building a $17 million headquarters in downtown Niagara Falls. Years before, the city had initiated an urban renewal project that had gone nowhere. It was hoped the new Hooker complex would spark life into the nearly desolate downtown from which custard stands, museums, and souvenir shops had gone, to be replaced by empty lots and an unsuccessful convention center. So anxious were officials to receive the new building that they and the state granted the company highly lucrative tax and loan incentives and gave it a prime parcel of property nearest the most popular tourist park on the American side, forcing a hotel owner to vacate the premises in the process.

Industry had begun its grip on the river as early as the mid-1700s, when Daniel Joncaire constructed a lumber mill just above the American falls, employing a system of overshoot wheels and pulleys to make practical use of the river's swift flow. Ventures like Joncaire's proliferated for the next hundred years, and included the excavation of another canal from the upper river to the lower to create additional power for flour-mill operations. There was concern at the time that sucking in water from above the falls and diverting it around the cataracts would lessen the flow to the point where it would detract from the beauty of the falls, but the Niagara Falls Hydraulic Company dispelled the notion: "Its attractiveness as a watering place will continue undiminished; for the proposed situation of the factories is such as to preclude the possibility of their detracting in the least from the grandeur of the cataract." Electricity was first produced from river power in 1881. Industries quickly filtered into the city, supplied by what was called the Niagara Falls Power Company at the turn of the century. Within a short span, aluminum, calcium chloride, ferroalloys, and other products were being manufactured quite economically because of the availability of cheap hydroelectricity.

The city had made the decision to accommodate industry at the expense of its great natural attractions. Tourists who ventured into town had to pass a two-mile row of unsightly factories before arriving at the key vantage points near the falls, and when they did, they could see streams of brownish suds in the turbulent waters at the base of the cataracts. Fishing had been largely destroyed in both the river and Lake Ontario, and parks and beaches, once of scenic value, had deteriorated so greatly that it was only on the hottest days they drew large gatherings. The smell of dead fish and garbage often permeated the winds, and the rapids of the river were now at lower levels than ever before—the electrical generators were sucking in too much water. Mink and deer that had once foraged in the brush were unable to reproduce as they formerly had, and the populations of muskrat and ringneck pheasant were dwindling at a rapid rate. The fertility of mallard duck eggs was less than half what it had been in previous decades, and other birds laid eggs with shells so brittle that they cracked from the slightest impact. The wildlife problems coincided strikingly with the increased volumes of chlorinated compounds being produced on the Niagara Frontier.

Despite the frightening environmental indications, O'Hara and the mayor, Michael O'Laughlin, continued to cater to industrial whims and to ignore those who might cause trouble for the plants. When residents appeared at city council meetings, O'Laughlin cut them short in their complaints during the public sessions. Karen Schroeder, for one, had great difficulty reaching the mayor or O'Hara on the phone to tell them of her distress. At one meeting, she said, Tim was told by a councilman, Pierre Tangent, that it was difficult for the city to attack the Hooker Chemical Company while negotiations for its new building were in progress. Obviously, a city-initiated lawsuit against the firm would have been quite untimely.

At the very time City Manager O'Hara was explaining to me that the Love Canal was not threatening human lives, both he and other authorities were aware of the nature of Hooker's chemicals. In the privacy of his office O'Hara, after receiving a report on the chemical tests at the canal, had discussed with Hooker the fact that it was an extremely serious problem. Even earlier, in 1976, the New York State Department of Environmental Conservation had been made aware that dangerous compounds were present in the basement sump pump of at least one 97th Street home, and soon after, its own testing had revealed that highly injurious halogenated hydrocarbons were flowing from the canal into adjoining sewers. Among these were the notorious PCBs—polychlorinated biphenyls. The Hudson River had become so badly polluted with these compounds that a $200 million project was initiated to dredge contaminated river sediments. PCBs, which are known to kill even microscopic plants and animals, also poisoned animal feed in at least seventeen states during 1979, leading to the destruction of millions of chickens and eggs from Oregon to New Jersey. Quantities as low as 1 part of PCBs to a million parts of normal water are enough to create serious environmental concern; in the sewers of Niagara Falls, the quantities of halogenated compounds were thousands of times higher. The other materials tracked in sump pumps or sewers were just as toxic as PCBs, or more so. Prime among the more hazardous ones was residue from hexachlorocyclopentadiene, C-56 for short. Few industrial products approach the toxicity of C-56, which was deployed as an intermediate in the manufacture of several pesticides whose use had created well-known environmental crises across the nation. The chemical is capable of causing damage to every organ in the body.

While the mere presence of C-56, however small the quantities, should have been cause for alarm, government remained inactive. It was not until early 1978—a full eighteen months after C-56 was first detected—that air testing was conducted in basements along 97th and 99th streets to see if the chemicals had vaporized off the sump pumps and walls and were present in the household air. The United States Environmental Protection Agency conducted these tests at the urging of the local congressman, John La Falce, the only politician willing to approach the problem with the seriousness it deserved.

While the basement tests were in progress, the spring rains arrived, further worsening the situation at the canal. Heavier fumes rose above the barrels. More than before, the residents were suffering from headaches, respiratory discomforts, and skin ailments. Many of them felt constantly fatigued and irritable, and the children had reddened eyes. Tim Schroeder developed a rash along the back of his legs and often found it difficult to stay awake. Another Schroeder daughter, Laurie, seemed to be losing some of her hair. Karen could not rid herself of throbbing pains in her head. Yet the Schroeders stayed on.

Three months passed before I was able to learn what the EPA testing had shown. When I did, the gravity of the situation immediately became clear: benzene, a known cancer-causing agent in humans, had been readily detected in the household air up and down the streets. A widely used solvent, benzene in chronic-exposure cases is known to cause headaches, fatigue, loss of weight, and dizziness at the onset, and later, pallor, nosebleeds, and damage to the bone marrow.

There was no public announcement of the benzene hazard. Instead, it seemed that some officials were trying to conceal the finding until they could agree among themselves on how to present it. Indeed, as early as October 18, 1977, Lawrence R. Moriarty, an EPA regional official in Rochester, had sent to the agency's toxic substances coordinator a lengthy memorandum stating that "serious thought should be given to the purchases of some or all of the homes affected. . . . This would minimize complaints and prevent further exposure to people." There was concern, he said, "for the safety of some 40 or 50 homeowners and their families." In an unsuccessful effort to learn the test results, I had regularly called the EPA and other sources, including the private laboratory contracted to conduct the tests; nervousness frequently crept into these discussions. No one wanted to talk.

Up until the second week of May 1978, I was still being told that the results were not ready, so I was surprised that same week to read a memorandum that had been sent from the EPA's regional office to Congressman La Falce. Buried in the letter was a sentence that referred to the analyses, saying that they suggested "a serious threat to health and welfare."

Immediately, local officials grew upset that the results had been publicly released. After an article of mine on the benzene hazard appeared in the *Niagara Gazette*, I received a telephone call from Lloyd Paterson, a state senator representing the Niagara area at the time.

"That Love Canal story," he began. "You know, you can panic people with things like that."

I explained to the senator that the finding was newsworthy and therefore it was my duty to print it. He responded that he did not want to see "people screaming in the streets." Irritation filled his voice as he continued: "We had a meeting last week, and there was no specific agreement on when this would be released."

The county health commissioner, Dr. Clifford, seemed unconcerned that benzene had been detected in the air many people were constantly breathing. There was no reason to believe their health was imperiled, he said. "For all we know, the federal limits could be six times too high," he stated with striking nonchalance. "I look at EPA's track record and notice they have to err on the right side." City Manager O'Hara, when I spoke to him in his office about the situation, told me I was overreacting to the various findings. He claimed the chemicals in the air posed no more risk than smoking a couple of cigarettes a day.

Dr. Clifford's health department refused to conduct a formal study of the people's health, despite the air-monitoring results. His department made a perfunctory call at the 99th Street Elementary School, and when it learned that classroom attendance was normal, it ceased to worry about the situation. For this reason, and because of growing resistance among the local authorities, I went to the southern end of 99th Street to make an informal health survey of my own. A meeting was arranged with six neighbors, all of them instructed beforehand to list the illnesses they knew of on their block, with names and ages specified, for presentation at the session.

The residents' list was startling. Either they were exaggerating the illnesses, or the chemicals had already taken an impressive and disheartening toll. Many people, unafflicted before they moved there, were now plagued with ear infections, nervous disorders, rashes, and headaches. One young man, James Gizzarelli, said he had missed four months of work because he had breathing troubles. His wife had experienced epilepsy-like seizures that she said her doctor was unable to explain. Meanwhile, freshly applied paint was inexplicably peeling from the exterior of their house. Pets too were suffering, most seriously if they had been penned in the backyards nearest the canal; they lost their fur, exhibited skin lesions, and, at quite early ages, developed internal tumors. There seemed also to be many cases of cancer among the women. Deafness was prevalent: on both 97th and 99th streets, traffic signs warned the passing motorists to watch out for deaf children playing near the road.

One 97th Street resident, a woman named Rosalee Janese displayed a number of especially suspicious symptoms. She lived at the canal's southern end, where the chemicals were leaking fastest and surface deterioration was most pronounced. Pimples and sores on her feet, arms, and hands caused her constant pain, and she suffered from daily bouts of nausea, faintness, internal pains, and a thick and oddly colored perspiration. The symptoms had begun suddenly, within weeks after a routine cleaning of her family's in-ground pool. During that chore, Mrs. Janese had been forced to hurriedly grab a rag and stuff it in the bottom drain: a black sludge was oozing through it into the pool.

Evidence continued to mount that a large group of

people, all of the hundred families immediately by the canal and perhaps many more, were in imminent danger. While they watched television, gardened, or did the wash, even while they slept, they were inhaling a mixture of damaging chemicals. Their hours of exposure were far longer than those of a chemical factory worker, and they wore no respirators or goggles. Nor could they simply walk out of the door and escape. Helplessness and despair were the main responses to the blackened craters and scattered cinders behind their backyards. But public officials often characterized the residents as hypochondriacs, as if to imply it was they who were at fault.

Timothy Schroeder looked out over his back land and shook his head. "They're not going to help us one damn bit," he said, throwing a rock into a puddle coated with a film of oily blue. "No way." His calls to the city remained unanswered while his shrubs continued to die. Sheri needed expensive medical care, and he was afraid there would be a point where he could no longer afford to provide it. A heavy man with a round stomach and a gentle voice, he had always struck me as easy-going and calm, ever ready with a joke and a smile. That was changing now. His face—the staring eyes, the tightness of lips and cheeks—candidly revealed his utter disgust. Every government agency had been called on the phone or sent pleas for help, but none of them offered aid.

For his part, Commissioner Clifford expressed irritation at my printed reports of illness, and there was disagreement in the newsroom on how the stories should be printed. "There's a high rate of cancer among my friends," he argued. "It doesn't mean anything." While it was true that the accounts of illness which I printed regularly were anecdotal, Dr. Clifford had even fewer grounds for an evaluation: Mrs. Schroeder said he had not visited her home, and neither could she remember his seeing the black liquids collecting in the basements. Nor had Dr. Clifford even properly followed an order from the state commissioner to cover exposed chemicals, erect a fence around the site, and ventilate the contaminated basements. Instead, he had arranged to have two $15 window fans installed in the two most polluted basements and a thin wood snow fence erected that was broken within days and did not cover the entire canal. When I wrote an article on a man who had contracted Hodgkin's disease at thirty-three, after a childhood spent swimming in the canal, Dr. Clifford telephoned me. He was brief: "When," he asked, "are you going to go back to being a reporter?"

Partly as a result of the county's inadequate response and pressure from La Falce, the state finally announced in May 1978 that it intended to conduct a health study at the southern end. Blood samples would be taken to see if there were any unusual enzyme levels indicating liver destruction, and extensive medical questionnaires were to be answered by each of the families. Hearing this, the many residents who had maintained silence, who had scoffed at the idea that buried chemicals could hurt them, began to ask questions among themselves. Would their wives have trouble bearing babies? Would their developing children be prone to abnormalities? Twenty years from now, would the chemicals trigger cancer in their own bodies?

As interest in the small community increased, further revelations shook the neighborhood. In addition to the benzene, as many as eighty other compounds had been discovered in the makeshift dump, at least ten of them potential carcinogens. The possible physiological effects were profound and diverse. Fourteen of the compounds could affect the brain and central nervous system. Two of them, carbon tetrachloride and chlorobenzene, could readily cause narcosis or anesthesia. Many others were known to cause headaches, seizures, loss of hair, anemia, and skin rashes. When combined, the compounds were capable of inflicting innumerable illnesses, and no one knew what different concoctions were being mixed underground. But even then no one realized, since only Hooker could know, that beyond the pesticides and solvents, far beyond the fly ash, one hundred additional chemicals would be identified during the next year, including one recognized by the state health laboratories as the most toxic substance ever synthesized by man.

In June, I learned from a former city bulldozer operator who had backfilled the canal that Hooker was not alone in its dumping. On three occasions, he claimed, men in army uniforms had pulled up to the dump in a jeep and truck to unload beer-barrel-shaped drums coated with what appeared to be zinc or lead. One of the men, a captain, had ordered that the containers be pushed gently into the deepest crevices of the former waterway. Were there radioactive substances in those barrels, or chemical-warfare residues? The United States Defense Department was never to answer. Nearby residents were forced to live with yet another unnerving uncertainty.

Edwin and Aileen Voorhees, as it turns out, had the most to be concerned about. When a state biophysicist analyzed the air content of their basement, he determined that the safe exposure time there was less than 2.4 minutes —the chemical content of the air in the basement was thousands of times the acceptable limit for twenty-four-hour breathing. This did not mean they would necessarily become permanently ill, but their chances of contracting cancer, for example, had been measurably increased. In July, I visited Mrs. Voorhees for further discussion of her problems, and as we sat in the kitchen drinking coffee, the industrial odors were easy to notice. I saw that Aileen, usually chipper and feisty, was filled with anxiety. She stared down at the table and talked only in a lowered voice. Everything now looked different to her. The home she and Edwin had built had become a jail cell. Their yard was just a pathway through which the toxicants entered the cellar walls. The field out back, that proposed "park," portended to be the ruin of their lives. I reached for her phone and called Robert Matthews, a city engineer who had been given the job of overseeing the situation. Was the remedial program, now in the talking stage for more than a year, ready to begin soon? No. Had there been any progress in deciding who would pay for it? No. Was there any chance of evacuating the Voorheeses? Probably not, he said: that would open up a can of worms—create a panic.

On July 14 I received a call from the state health department with some rather shocking news. The preliminary review of the health questionnaires was complete, and it showed that women living at the southern end of the canal had suffered a high rate of miscarriage and given birth to an abnormally large number of children with birth defects. In one female age group, 35.3 percent had records of spontaneous abortion. That was far in excess of the norm: the odds against its happening by chance are 250 to 1. Four children in one small section of the neighborhood had documentable birth defects, clubfeet, retardation, and deafness. These tallies, it was stressed, were "conservative" figures. The people who had lived there longest suffered the highest rates.

I shuddered when the health officials, with their statistics and microscopes, their access to confidential medical records, confirmed what I had feared for months. Returning from the newspaper office, I thought of a recent trip to the dumpsite when a young boy had approached me on his bicycle while I inspected a surfaced barrel. He climbed off the bike and stood there in silence, blankly staring at me.

"Hello," I said.

He remained silent.

"Do you know of anywhere else where drums have popped up like this on the field?" I asked.

He did not respond at first. Finally, gurgling his words so badly that I could not understand him, he pointed to his left ear. His motions told me he could not hear—he was deaf or retarded. Then he mounted his bicycle and made his way to the south end of 97th Street.

I called the Schroeders to tell them the news from the state. They too found it upsetting. As Karen spoke, her husky voice quavered. We spoke of Sheri and her numerous birth defects. Now it seemed more likely than ever that it was the chemicals in their backyard that had so severely affected the girl's health. When Karen became too unnerved to speak further, Tim took the phone. He was confused: a city official had talked to him that day, after attending the meeting with state officials at which the medical data were discussed, yet he apparently had told Tim there was nothing new to report. That infuriated me. Hanging up the phone, I felt a tightness grow at the pit of my stomach until it reached the point of nausea. For too many months I had been a bystander to an accident occurring in slow motion, watching people suffer while the government machinery slipped and ground to frequent halts, to the serious detriment of helpless adults and children. How many more would become sick from additional exposure while the local officials fumbled the issue and attempted to cover it up? How many of the children would have had normal lives had Hooker warned of the dangers?

The data on miscarriages and birth defects, coupled with the other accounts of illness, finally pushed the state's hierarchy into motion. A meeting was scheduled for August 2, at which time the state health commissioner, Dr. Robert Whalen, would formally address the issue. The day before the meeting, Dr. Nicholas Vianna, a state epidemiologist, told me that it also appeared that residents were incurring some degree of liver damage: blood analyses had shown hepatitis-like symptoms among the enzyme levels. Dozens if not hundreds of people had been adversely affected. In the meantime, Donald McNeil, a most perceptive reporter for the *New York Times*, arrived in town. I gave him my Love Canal file and a list of phone numbers, and invited him to be my house guest in order to explain to him more fully the background of the unusual situation.

Before leaving for the state capital, Albany, for the Whalen address, I was scheduled to attend a meeting with executives at Hooker. Increasingly nervous about what they were reading in the newspaper, and equally annoyed by it, they had decided to break their prolonged silence and answer some questions about the canal. But I was not to face them alone; my publisher, Susan Clark, was also asked to attend. This fit a pattern of theirs: it seemed that when they wished to soften my reporting, they would go to her directly and find a more sympathetic ear.

In Hooker's lavish boardroom were gathered public relations men, an engineer, a company physician, and a toxicologist. As we entered, Ms. Clark skirted over to Charles Y. Cain, the company's public affairs vice-president, and hugged him. This made me nervous. The meeting had been called because Cain seemed unhappy with my reporting, and I looked upon the session as an austere occasion at which I would attempt to pose some difficult questions and crack the shell of Hooker's secrecy.

Despite Cain's initial friendliness, tension soon crept into the room. Cain sat down at the head of a long wood table, took off his suit jacket, and looking me grimly in the eye, began the session on a discordant note.

"Let me just say this," he said. "We just did a little survey, and we found Hooker is more popular in town than the *Niagara Gazette*."

I was taken aback at his opening remarks and decided to make light of them. "Well, we did a little checking too," I said, "and we found the most popular business in town is neither Hooker nor the *Gazette*, but Kach's Korner." The reference was to a topless-dancer tavern near their main plant.

"And your mother wears army boots," Cain responded.

For the next three hours, the Hooker scientists and executives, in more diplomatic tones, attempted to undercut some of my reporting. They charged that I had employed the word "toxic" too liberally, and their physician, Dr. Mitchell Zavon, said he did not believe any of the people at the dumpsite, not even Mrs. Voorhees, to be in any form of danger. Hooker was acting, and had acted, as a "responsible corporate citizen," they repeatedly claimed. The canal, in their view, was the "best available technology" for waste disposal at the time, and they thought the waterway constituted a "sealed clay vault." Although they had no current legal obligation to do so, they emphasized that they had cooperated with the city and school board in paying for an engineering study.

But what of the people? Did Hooker honestly believe they weren't in trouble? Did those at the meeting honestly believe, despite their expertise in chemistry and biology, that their waste material was not toxic?

Dusk was settling as we left the Hooker offices. I drove through downtown, where the tourists had once cluttered Falls Street, and thought about the city. The spray from the falls was still visible above the treetops of Prospect

Point. On a chilly evening, when a temperature inversion stopped the breeze, the factory excretions would appear almost frozen above their smokestacks, against the blackened backdrop. Above the swift Niagara, there would be a cold, intimidating dark void. Born and raised in this city, I had seen these sights all my life. Yet there was a foreign feeling about it all now.

When I arrived at the government complex in Albany the next morning, I rode to the fourteenth-floor health offices with Dr. Clifford, who made joking references to the local newspaper's story on possible evacuations at the dump. He found that quite absurd. I entered the department's public relations office and picked up a copy of the *New York Times*. There on the front page, in the lower left corner, was the story: "Upstate Waste Site May Endanger Lives."

Dr. Whalen had begun his meeting just before I arrived. As I entered the auditorium where he was speaking, I spotted the Schroeders and Aileen Voorhees in the small audience, watching intently. But not Tim; he was bent over in his chair, staring at the floor. None of them expected anything to happen, but they had traveled the five hours to Albany anyway, to make their presence felt.

Minutes later, to their surprise, Dr. Whalen read a lengthy statement in which he urged that pregnant women and children under two years of age leave the southern end of the dumpsite immediately. He declared the Love Canal an official emergency, citing it as a "great and imminent peril to the health of the general public."

2

"YOU'RE LETTING US DIE"

When Commissioner Whalen's words hit 97th and 99th streets, by way of one of the largest banner headlines in the *Niagara Gazette*'s 125-year history, there ensued a tumult among the residents that approached a state of panic. Returning from Albany that night, I was told that dozens of people had massed on the streets, shouting into bullhorns and microphones to voice frustrations that had been quietly accumulating for months. Many of them vowed a tax strike because their homes were rendered unmarketable and unsafe. They attacked the government for ignoring their welfare. Previously, their emotions had been contained to the point of apathy, with only the Schroeders and several of their immediate neighbors speaking their minds. But now a man of high authority, a physician with a title, had confirmed that their lives were in danger. Most wanted to leave the neighborhood immediately.

Terror and anger roiled together, exacerbated by Dr. Whalen's lack of provision for a government-funded evacuation plan. His words were only a recommendation: it was up to individual families to decide if they would risk their health and remain or abandon their houses and, in so doing, write off a lifetime of work and savings. The residents were also outraged that they and their older children were not addressed in the commissioner's statement.

Late into the evening they served as guides for the retinue of reporters that had flocked into town that day, taking the journalists onto the Love Canal and thrusting sticks into rotted chemical drums to display the acrid contents.

On August 3, with the media's klieg lights barraging the neighborhood, Dr. Whalen decided he should speak directly to the people. He arrived with a deputy, Dr. David Axelrod, who had directed the state's investigation, and also with Thomas Frey, a key aide to Governor Hugh Carey. Before his plane left the state capital, I received a call from another assistant, Dr. Stephen Kim, inquiring if the mood of the people was such that it would be prudent to enlist police protection for the officials. I told him it would be appropriate indeed. Not without reason, the residents felt their lives hung in the balance, and their demeanor threatened to turn ugly.

When I arrived that night for a public meeting in the 99th Street School auditorium, I was glad I had made the suggestion on security. There was a wave of fright and hostility rolling from one end of the school to the other. The residents stood in small groups, awaiting the officials' arrival with visible tension. Many eyes were red, faces were bewildered and downcast, as they watched the health researchers file in for what was later to be known as the "hell meeting." Suddenly a man who looked about forty years old ran up to an official, pleading that his children be moved away. When no assurances came, he threw himself onto the linoleum floor of the hallway, rolling over and over in a fit of hysterical crying. Young mothers sat in the auditorium, children on their laps, and openly wept. Others looked blankly toward the podium, dazed.

Frey was given the grueling task of controlling the crowd of five hundred people. In an attempt to calm them, he immediately announced that a meeting between the state and the White House had been scheduled for the following week. The purpose would be to classify the Love Canal as a national disaster, thereby freeing federal funds. For now, however, he could promise no more. Neither could Dr. Whalen and his staff of experts. All they could say was what, by now, was already known: twenty-five organic compounds, some of them capable of causing cancer, had invaded the residents' homes, and because young children were especially subject to toxic effects, they should be moved to another area.

The emotional assemblage was unimpressed. For the next four hours, the meeting was a shouting match between the people and their government. They screamed and shook their fists in the air. "You're letting us die, dammit!" shouted a man from the rear of the room. "You're going to stand there and watch us all die!"

As I looked around the frantic gathering, I noted that nowhere in the building were there any members of the city council. The mayor was away on vacation in Ireland, and had decided not to return for the crisis. City Manager Donald O'Hara was there in the mayor's place. His entrance was greeted by raucous catcalls and profanity.

The hot, humid air inside the auditorium now reverberated to the voice of Thomas Heisner, a lanky man in his thirties whose intense face sparkled with beads of perspiration. A nextdoor neighbor of the Schroeders, he too had fought a long and fruitless battle against the dump's encroachment on his land. There were times, after a rainfall, when liquids of various colors streamed down his driveway to the gutters of the street. Heisner stood near the back of the room and held both arms high above his head for attention. "Someone once said, 'Give me liberty or give me death,' " he shouted. "Now I'm telling you, I'm telling you we're not backing down to anybody.

We're not giving up. My child has a congenital birth defect, and we're going to fight the stinking Hooker chemicals. You can bet on that. We want to live and we're going to fight."

Heisner's pretty wife, Florence, stood up after him, crying, "How do you determine if a child should stay? Do I let my three-year-old stay? Where do you draw the line?" Then she pointed in the direction of the canal: "Whatever's in there is in us now."

Dr. Whalen's order had applied only to those living at the canal's southern end, on its immediate periphery. From previous research, it seemed to me that the problem was larger than that. Families living across the street from the dumpsite, or at the northern end where the chemicals were not so visible at the surface, suffered afflictions remarkably similar to those of families whose yards abutted the southern end. I interviewed dozens of additional people in the school hallway, and determined that, indeed, effects of the canal were probably frighteningly widespread. Serious respiratory problems, nervous disorders, and rectal bleeding were reported by many who were not covered by the order. During the summer months, with windows open and the fumes clinging close to the ground, illnesses seemed to increase markedly. I returned to the news office and, extracting from endless pages of health accounts, wrote a story on what I had found.

Throughout the following day, residents posted signs of protest on their front fences or porch posts: "Love Canal Kills" or "Give me liberty, I've got death." Emotionally exhausted and uncertain about their future, men stayed home from work, congregating on the streets or comforting their wives as they cried. By this time the board of education had announced the closing of the 99th Street School for the following school year because of its proximity to the exposed toxicants. Still, there was no public relief for the residents.

Another meeting was held that evening, at a fire hall on 102nd Street. It too was unruly, but the people who had called the session in an effort to organize themselves managed to form an alliance, the "Love Canal Homeowners Association," and elect as president Lois Gibbs, a pretty woman with jet-black hair whose age, twenty-seven, belied what proved to be a formidable ability to deal with experienced politicians and to keep the matter in the news. She lived two blocks east of the canal, on 101st Street. After Mrs. Gibbs's election, Congressman John La Falce entered the hall and announced, to wild applause, that the Federal Disaster Assistance Administration would be represented the next morning, and that the state's two senators, Daniel P. Moynihan and Jacob Javits, were working with him in an attempt to get funds from Congress. "Up until now, I had lost the American vision and dream," said Thomas Heisner, sobbing slightly at the microphone. "The city ruined that vision. Now my heart is happy." Meanwhile, members of the local media pledged grants to temporarily relocate pregnant women and young children. Simultaneously, the state announced that it would pay apartment rent for thirty-seven families who had pregnant women or children under two in their households. Attention still focused on the southern end of the canal. The Schroeders began packing for a move to an old air-force-base apartment complex, where they would soon be joined by many of their neighbors. Other families filled the downtown hotel rooms, aided by community donations of money and food. One of those to contribute was the Frank Gannett Newspaper Foundation, formed by the newspaper chain of the same name. It donated $6,000 to the state's multi-million-dollar rescue effort—and then gave its donation front-page treatment for several days running in the *Gazette*. There was a measure of irony in this, for the newspaper's publisher, Susan Clark, had hardly encouraged my pursuit of Hooker's activities, and the news chain itself had generally placed such a restrictive budget upon the news that it was more of a hindrance than a help in my reporting.

The residents at the northern end continued to concern me. The first indication there of significant chemical migration had come the previous July, when I visited the home of Russell Taylor on 97th Street, more than half a mile from the "crisis" zone. Taylor told me that as he was breaking a hole in his basement floor to install a bathroom, a syrupy red fluid suddenly filled up the hole. He filled two garbage pails with the strange matter, but before he could take them from the house his young collie stuck its nose into one of them. The dog developed severe nasal problems as a result, cancerlike lesions that grew so painful they had to put it away.

At my suggestion, the Taylors had collected some of the fluid from their sump pump in a glass jar. I took the sample to a local laboratory equipped with the tools to detect chlorinated hydrocarbons and other organic chemicals. The results arrived weeks later, on August 4. The chemists at the laboratory were dismayed by what they had found. "As soon as we opened the lid, there was that C-56 smell," said one. "And we found it in the tests. We also found some PCBs. I just get the shakes thinking of those people there."

The people at the north end were apprehensive and discouraged. They did not wish to leave their homes, but neither did they appreciate the lack of attention they were experiencing. They knew they had the same problems as the south end. Before my arrival, the state had quietly handed a number of them slips of white paper that listed the chemical compounds found in their basement air. It was clear that there was extensive contamination with benzene, chloroform, and toluene, in several cases to a degree that approached the readings at the south end.

More disturbing facts about the extent of contamination continued to accumulate. From the slopes of the terrain and the low points where creekbeds and swales had been filled, there were indications that chemicals had long since traveled outside the channel's banks, farther even than the first two "rings" of homes alongside the dump. Nearly a mile to the north of the Schroeder home, I noticed one such downgrade of land near a small, neat house with the nameplate "Moshers" hung on a post in the front yard. I knocked on the door, announced my purpose, and was allowed to enter by a thin, pale man who received me with reluctance. We walked to the kitchen to meet his wife, Velma, a fifty-four-year-old woman confined to a wheelchair and barely able to speak. She too was wan and fragile. "I'm just so tired all the time," she explained. "I'm just so tired, and I don't think they know what's really wrong with me." She said her great fatigue had set in more than a dozen years before, at a point in her life when she was operating a beauty shop in her basement. "It didn't smell right down there," she added. "Not at all. I'd get headaches all the time. I would go out back at night, to play croquet, and my legs would give way, just collapse." She had closed the salon when she was no longer able to navigate the stairs.

Mr. Mosher was not as candid as his wife. When I asked about his health, he stepped back from me as if I had uttered a blasphemy. His reaction, I soon learned, was out of fear that any publicity would affect his standing at a local carbon plant, where he held a managerial position.

I walked toward the door leading to the basement. "Do you have a flashlight?" I asked.

Mr. Mosher nodded his head and returned with one promptly. As we descended the stairs, he explained to me that no one had checked his home for contamination, so he had not worried about it. I stirred the sump-pump sediment with a piece of wood and switched on the flashlight. There it was: a red, rubbery substance like that described by Russell Taylor.

Having encountered that, I grew impatient with Mr. Mosher's reticence about his own health and warned him that it could be endangered. Having seen, in the sludge of the sump pump, that chemicals might have found a way into his cellar, he said, "Well, I've got some heart problems. And I had an enlarged spleen removed. It was twelve and a half pounds."

Velma, who had heard our conversation, began to speak of the summer nights when strong fumes from the canal rendered their bedroom such a trap container for pungent air that breathing was made difficult, and sleep even more so. Then the woman weakly cocked her head to one side and stared up at her husband. "Tell him about your problem," she insisted.

Mr. Mosher stood where the hallway met the kitchen and stared at the floor. After a minute's silence, he looked up at me. In a low tone he said, "I've got cancer, in the bone marrow. They're treating me for it now."

Upon returning to the office, I searched through a book on toxicology, *Dangerous Properties of Industrial Materials*, for the symptoms of benzene poisoning. There was a lengthy list that included fatigue, edema, narcosis, anemia, and hypoplastic or hyperplastic damage to the bone marrow. Although the hour was approaching midnight on a Sunday, I felt compelled to call Dr. Axelrod of the state health department to inform him of the Moshers' condition. He told me that not far from their home, researchers from his unit had indeed detected benzene in the air.

The implications of what I had seen once again began to take their toll on me. Fatigued after the long day, I lay down on a desk in the center of the room and attempted to fall asleep. But each time I closed my eyes, I saw in my mind the haunting image of the Moshers, a devastated couple, speaking to me in their kitchen. Giving up the idea of sleep, I began to write an article for the next day's newspaper on the spread of the chemicals far away from the current crisis area.

The ever-deepening official uncertainty and disorganization in developing a rescue plan had a marked effect on the attitude of the people and on their psychological stability. Unable to cope with their jobs, men remained at home. Children, exposed to an inundation of television reports or radio messages on their danger, began to have nightmares of death. At the elementary school, which had been quickly converted into a crisis center, hundreds of residents arrived now from each end of the canal and from streets beyond, jamming the hallways as they waited in long lines to have their blood drawn and tested, or to sign up for temporary evacuation or other aid. Women and children wailed as they waited their turn. Under such pressures, many marriages showed signs of strain, for while the women often wanted to leave immediately, the men were not inclined to do so without compensation for their homes. Bitter arguments also divided the Love Canal Homeowners Association. Those who lived immediately near the chemicals believed there would be only a limited amount of government money forthcoming, and they demanded that they be the first to go. They resented having homeowners as far away as 103rd Street request evacuation, and to fight for the largest portion of the funding, they formed a radical splinter group.

Sensing that circumstances were developing out of control, the governor's office announced that Hugh Carey would be at the school on August 7 to address the people. In an election year, with the public keenly tuned to a story now dominating the national media, decisions were being made in Albany and Washington. Hours before the governor's arrival, a sudden burst of reports labeled "urgent" came across the newswires from Washington. President Jimmy Carter had officially declared the Hooker dumpsite a national emergency, analogous to a tornado or a flood.

Hugh Carey's arrival was met with applause from the weary people. The governor announced that the state, through its Urban Development Corporation, planned to purchase, at fair market value, those homes rendered uninhabitable by the marauding chemicals. He spared no promises: "You will not have to make mortgage payments on homes you don't want or cannot occupy. Don't worry about the banks. The state will take care of them." By the standards of Niagara Falls, where the real estate market was depressed, the houses were in the middle-income range, worth from $20,000 to $50,000 apiece. The state would assess each house and purchase it, and also pay the costs of moving, of temporary housing during the transition period, and of special items not covered by the usual real estate assessment, such as installation of telephones. In all, the state would eventually pay well in excess of $30 million for its investigation of the problem, the purchase of homes, temporary accommodation, and the remedial drainage program which Carey pledged would begin as soon as possible, to stop the further migration of toxicants outside the dumpsite. Federal money would also go toward the remedial construction, but in amounts that were only a small percentage of over-all costs.

But what the governor, in his surprising generosity, did not say was how many homes would be deemed "uninhabitable." The fear persisted among those in the outer rings that they would be left behind, with benzene and toluene infiltrating their basements, and their property, located near a wasteland, quite literally of no value whatsoever. At first it was decided that the state would buy the "ring one" homes adjacent to the waste pit on 97th and 99th streets. This meant that those who lived across from the Schroeders, in "ring two," would remain where they were, with only ninety-seven homes taken over by the Urban Development Corporation. There were problems with such a limited rescue mission. Chemicals did not respect the boundaries of asphalt roads, did not find them a barrier to their progressive movement, and it seemed likely that underground water had carried the contaminants through breaches in the clay or sand and gravel lenses to the opposite sides of 97th and 99th streets, perhaps in large quantities. When I reviewed the basement-air samplings conducted in "ring two," I found that in certain cases the levels of the same type of compounds—solvents such as toluene, trichloroethane, and benzene—were higher than on the closer streets. Some of the state people, confused by the readings and not wanting to create any further hysteria, told the homeowners that these levels might have originated from household items such as nail polish, paint thinners, and canned gasoline, which contain some of the same solvents found in the canal. However, it seemed much more likely that the canal's reach was farther and stronger than expected and that the state, not wanting its "zone of contamination" to be any larger than what it had already committed itself to, had begun making light of the additional migration, soft-pedaling the significance of its own tests. Families in these "ring two" homes exhibited many of the same health disorders found across the street, and others as well. On 97th Street, the family of Ronald Zuccari had two children with birth defects, among them a deficiency of tooth enamel. When I visited their nextdoor neighbor, I found a child there also with insufficient tooth enamel. Throughout the neighborhood were distressing stories of nosebleeds, asthma, ear problems, and miscarriages.

Within days after the "ring two" conditions were brought to light, Thomas Frey of the governor's office announced that those homes too would be purchased. That brought the total number of houses to be abandoned permanently to approximately 240—nearly 1 percent of the city's habitable structures would be lost to the realty market at a single stroke. There had been many earlier cases of large communities being evacuated because of chemical explosions, but this evacuation was to be permanent, and it was caused, not by a simple tank-car derailment, but by substances gradually pervading the soil. It had never happened before on such a scale.

Soon the state, coordinating the crisis through its health and transportation departments, began the awesome task of mass evacuation. Ironically, its offices were installed in the endangered 99th Street School, while the students were transferred to classrooms elsewhere in the city. Houses were appraised individually, and one by one, the homeowners were brought in by appointment to negotiate a settlement. Some residents, more worried about their bank accounts than about their health, refused to leave, causing an endless cycle of renegotiations until a compromise was reached. Many residents remained unconvinced that they had received what their homes were actually worth. As Aileen Voorhees later remembered, "They put us there in the Falcon Manor air base for three months, from August to November. We looked at so many homes it was ridiculous—every chance we got we went out to look, and it took us months. We couldn't find a brick house like we had, there was no way we could touch a brick one at the prices. It was so discouraging, you didn't know where to turn. You couldn't take it. You weren't a bit happy, thinking, 'You got to move, you got to move, but where are you going?' We had all the work done in our other house, painting, varnishing, the yard. We had to do it all over again. I still haven't gotten over being forced from my house."

First in a trickle and then, by September, in droves, the families gathered their belongings and carted them away. Moving vans crowded 97th and 99th streets. As if working on an assembly line, linesmen went from house to house disconnecting the telephones and electrical wires, while carpenters pounded plywood over the windows to keep out vandals. With slow inevitability, the nighttime house lights in the neighborhood eerily dwindled until, by the following spring, 237 families were gone, 170 of them to new homes. Several of them, including the Thomas Heisners, felt so attached to their houses that they had them lifted from the foundations and transported to a safer parcel of land. In time the state erected around the first two rings a green chain-link fence eight feet in height. There was now a clearly demarcated contamination zone comprising a six-block residential stretch.

The psychological impact of suddenly relocating was not to be readily measured, but certainly it was great. More than seven hundred people had learned, in a matter of

140 hours, that they had to pull up stakes forthwith and watch helplessly as their neighborhood disbanded. The long-term residents found it especially difficult to accept the tumultuous events. In her Irish accent, Mrs. Josephine Craig of 99th Street explained: "Moving was a great loss. I loved my house. I loved the neighborhood. I had a sense of desolation, horrible. It doesn't hit you at the time you move, but when you look out your window and it's not the same view you've had for fifteen years, when you realize the people you would meet walking down the street you will never see again, it was horrible. You know, if you're moving yourself, you don't have these same feelings. When you are forced, there is a feeling of numbness. My nerves give me a lot of trouble. I can't eat. I have a choking feeling in my throat, a knot. It all caught up on me. I thought I would be there the rest of my life, but I was completely uprooted. It is that sense of loss." In the course of the evacuation, several persons suffered nervous collapse, and I was told that one elderly woman, seriously depressed upon hearing the state's plans, had died of heart failure.

In October 1978, the long-awaited remedial drainage program began at the south end. Trees were grubbed up, fences and garages torn down, and swimming pools removed. So great was the remaining residents' fear that dangerous fumes would be released over the surrounding area that the state, at a cost of $500,000, placed seventy-five buses at the ready at emergency evacuation pickup spots during the months of work, in the event that outlying homes had to be quickly vacated because of an explosion. The plan was to lay drainage tiles around the periphery of the canal, where the backyards were once located, in order to divert leakage to 17-foot-deep wet wells from which contaminated groundwater could be drawn and treated by filtration through activated carbon. (Removing the chemicals themselves would have been financially prohibitive, perhaps costing as much as $100 million—and even then the materials would have had to be buried elsewhere.) After the trenching was complete and the sewers installed, the canal was to be covered by a sloping mound of clay and planted with grass. One day, it was hoped, the wasteland would become a park.

It was the city's job to hire the contractor. City Manager Donald O'Hara's staff chose for the work a powerful local waste disposal firm, Newco Chemical Waste Systems. This in itself was a significant decision, for Newco had a long history of business transactions with both Hooker and the municipality. The firm's principals were serving as the developers for the new Hooker headquarters, and Newco had for quite some time been under contract with the city for disposal of its wastewater treatment-plant sludges. Because the drainage program was deemed an emergency project, the city was not required to put it out for strictly competitive bids among local construction firms, and there were virtually no cost restraints. Under the lucrative arrangement, Newco was simply to bill the government for materials and man hours, whatever they might be. I was later to learn that the remedial program for the south end, about six acres of land, would cost ten times what had originally been estimated in 1977. It was now a multi-million-dollar project. The company valued its president's time at $100 an hour, its vice-president's at $95 an hour, and the services of lower-echelon managers and foremen at equally exorbitant rates. Not long after the work commenced, O'Hara resigned his city manager's job to take a position with a Newco business affiliate, a move that surprised some of his friends and, right or wrongly, raised the question of impropriety.

In spite of the corrective measures and the enormous

effort by the state health department, which took thousands of blood samples from past and current residents and made countless analyses of soil, water, and air, the full range of the effects remained unknown. In neighborhoods immediately outside the official zone of contamination, more than five hundred families were left behind near the desolate setting, as the state had announced that it would buy no more homes. Their health remained in jeopardy.

The first public indication that the chemical migration had probably reached streets to the east and west of 97th and 99th streets, and to the north and south as well, came on August 11, 1978, when sump-pump samples I had taken from 100th and 101st streets showed traces of a number of chemicals found in the canal itself. One of these was lindane, a restricted pesticide suspected since 1948 of causing cancer in laboratory animals.

While probing 100th Street, I had knocked on the door of Patricia Pino, a blond divorcee of thirty-four with a young son and daughter. Her situation was of immediate concern to me because I had noticed that some of the leaves on a large tree in front of her house exhibited a black oiliness like that on the trees and shrubs on 99th Street. Her home was located near what had been a drainage swale. After I had extracted a jar of sediment from her sump pump for analysis, we talked about her family and what the trauma now unfolding meant to them. Ms. Pino was extremely depressed and equally embittered. Both of her children had what appeared to be slight liver abnormalities, and the son had been plagued with "nonspecific allergies," teary eyes and sinus trouble that improved markedly when he was sent away from home. Although it seemed unlikely that the contaminants could have traveled that far in great enough quantities to do physical harm, Patricia told me of times during the heat of summer when fumes were readily noticeable in her basement and sometimes even upstairs. She herself had been treated for possible cancer of the cervix. But like the people of the inner rings shortly before, her family was trapped.

On September 24, 1978, I obtained a state memorandum that said chemical infiltration of the outer regions was significant indeed. The letter, sent from the state laboratories to the United States Environmental Protection Agency, said, "Preliminary analysis of soil samples demonstrates extensive migration of potentially toxic materials outside the immediate canal area." There it was, in the state's own words. Not long afterward, the state's medical investigator, Dr. Nicholas Vianna, told me there were indications that residents from 93rd to 103rd streets might also have incurred liver damage. Along with the pancreas, the liver, because of its many functions, is the key indicator of toxicological damage. In its versatility and indispensability it has no bodily equal. The largest glandular organ in the body, weighing about four pounds, the liver is charged with storing vitamins, minerals, and proteins, manufacturing bile for digestion, and releasing glucose, in carefully measured units, to keep blood sugar at normal levels. What is more relevant, it receives blood from the stomach and intestines in order to remove wastes and poisons from it. Certain chemicals that enter the liver react with its tissue and enzymes in a way that unbalances the system and causes alterations in the blood. Thus, signs that the livers of residents outside the evacuated neighborhood might have been affected served to show that the fenced boundaries of the "contaminated zone" were much too narrow.

On October 4 Jon Allen Kenny, a young boy who lived quite a distance north of the evacuation zone, died of kidney failure. As the kidneys are known to be readily affected by toxicants, suspicion arose that the boy's death was related in some way to a creek that flowed behind his house and carried, near an outfall, the odor of chlorinated compounds. Because the creek served as a catch basin for a portion of the Love Canal, the state studied an autopsy on the boy, but no conclusions were reached. Jon Allen's parents, Norman, a chemist, and Luella, a medical research assistant, became discontented with the state's investigation, which they felt was "superficial." Luella said, "He played in the creek all the time. There had been restrictions on the older boys, but he was the youngest and played with them when they were old enough to go to the creek. We let him do what the other boys did. He died of nephrosis. Proteins were passing through his urine. Well, in reading the literature, we discovered that chemicals can trigger this. There was no evidence of infection, which there should have been, and there was damage to his thymus and brain. He also had nosebleeds and headaches, and dry heaves. So our feeling for now is that chemicals probably triggered it."

Mrs. Kenny went on: "We never noticed any signs of illness before. On June 6, there was a swelling around his eyes. They said it was an allergy, but the antihistamines didn't work, and his stomach became distended. On the third visit to the doctor's, the protein was found in his urine. Immediately he went to the hospital, for three weeks. He came home. He was fine. Then, a relapse. By August 31 he was free of protein but was kept on medication. Even so, in mid-September he had another relapse and took an atypical reaction—convulsions. No one knew what caused it. He was unconscious for two days, then he died. He was seven." When a state biophysicist visited Mrs. Kenny, an interesting, if perhaps unrelated, coincidence occurred. While the scientist stood out back, a bird swooped down near the creek outfall and, before his eyes, collapsed and died.

The probability that water-borne chemicals had escaped from the canal's deteriorating bounds and were causing problems at a distance from the site was not lost upon the Love Canal Homeowners Association and its president, Lois Gibbs, who was attempting to have additional families relocated. Because she lived on 101st Street, she was herself one of those left behind, with no means of moving despite persistent medical difficulties in her six-year-old son, Michael, who went under the surgeon's knife twice for urethral strictures. Mrs. Gibbs's husband, a worker at a chemical plant, brought home only $150 a week, she told me, and when they had subtracted $90 a week for food and other necessities, clothing costs for their two children, $125 a month for mortgage payments and taxes, utility and phone expenses, and medical bills, there was hardly enough cash left to buy gas and cigarettes, let alone to vacate their house—only a year or so away from being paid for in full—and rent an apartment.

Assisted by two other stranded residents, Marie Pozniak and Grace McCoulf, and the professional analysis of a Buffalo scientist named Beverly Paigen, Lois Gibbs mapped out the swales and creekbeds, many of them long ago filled in, and set about interviewing those many people who lived on or near formerly wet ground. After this survey, they came to the conclusion that these people were suffering from an abnormal number of kidney and bladder troubles and disorders of the reproductive system. In a report to the state, Dr. Paigen claimed to have found, in 245 homes outside the evacuation zone, 34 miscarriages,

18 birth defects, 19 nervous breakdowns, 10 epilepsy cases, and high rates of hyperactivity and even suicide.

To appreciate what this meant, one might consider the case of Lynn Tolli and her family, on 102nd Street. They lived, and continue to live, in a small white house on the west side of the street, about a quarter-mile from the "ring one" homes. Lynn was healthy before moving into the house, but since then had developed epileptic seizures, as had one of her sons. Another son and her husband, Harry, had suffered from respiratory illnesses, and one year a cat and a dog had died of no clear cause on their basement floor. The source of trouble may have been a manhole in the backyard, for when it rained the sewer overflowed, with an odor that reminded Lynn of passing by the Hooker plant. Upon analyzing the air in her home, the state discovered that it was contaminated with benzene and toluene to the extent that they advised the Tollis not to linger in the basement and not to use one of the first-floor bedrooms. Eventually one son became so afflicted with asthma that, despite possible psychological complications, they sent him to live with his grandparents.

In its roundabout way, the state health department, after an elaborate if somewhat disorganized investigation, confirmed some of the homeowners' worst fears. On February 8, 1979, Dr. David Axelrod, who by now had been appointed health commissioner and whose excellence as a scientist was widely known, issued a new order that officially extended the health emergency of the previous August, citing high incidences of congenital deformities and miscarriages in areas that had formerly been swampy or where creeks and swales had once flowed. With that revelation, the state offered to temporarily evacuate families with pregnant women or children under the age of two from the outer rings, up to 103rd Street. Still, there would be no further purchases of homes, nor another massive evacuation, temporary or otherwise. Under the new plan, those who left would be forced to return when their children passed the age limit.

Eventually, twenty-three families took the state up on its offer. Another seven families, ineligible under the plan but with adequate financial means, simply left their homes and took the huge loss of investment. Soon boarded windows speckled the outlying neighborhoods, serving as a constant reminder of the potential dangers to those who, like Marie Pozniak, were not assisted in moving. Mrs. Pozniak, who lived immediately north of the canal and was a victim of cervical cancer at quite an early age, poignantly described her feelings to me. "Being left behind," she said, "is the most terrible experience, horrifying. You know the dangers are still there. You can't get away from them; we all have benzene in our homes. It's a desperate feeling, one of inadequacy. You go out and see someone carrying in groceries into their home, in another part of town, and they're smiling and happy and you resent them for it, because they have a safe home. Yours is dangerous. You're afraid of your own home. My family is going to pieces, and there are divorces all over this place. My husband is so desperate: There is no way he can get us out, no way for him to protect his family, and that gets to him. That gets to everybody. They [United Way of America] have set up counseling, and hopefully that will help some of them. Most of the doctors around here don't want to become involved. To be truthful, I don't think they know enough about chemicals—they are totally uninformed, and they don't want to be informed. My doctor told me to relocate my daughter, even if it means separating her from us. She has asthma. She doesn't want to go. I sent her away for Easter, she was so congested, and just one day away and she didn't need her medication. The state knows this. My children are so upset—we can't have a garden this year. And they're afraid to go down into the basement. I know a woman who had to take her child for counseling. The kid was afraid of dying. Many of them are. They are thinking far beyond their years. She had to go too. She's afraid her children will die."

There were many signs of psychological distress. I watched wives move away from their husbands as the Love Canal became an obsession. Children watching television would see news reports on the contamination being read against a graphic backdrop of skulls and bones. Terrified that the Love Canal would come and get them like a bogyman in the night, they hid under couches and beds and would not come out. As the stress grew more intense, some of the parents sent their children to their relatives, even though in certain cases that meant hundreds of miles away. They did not wish their young to be the first human coalmine canaries.

Adding to the anxiety was a series of events that had occurred the previous November and December, not long after the evacuation of 97th and 99th streets. With increased attention to the canal's seepages, I had become interested in the possibility that Hooker might have buried in the Love Canal waste residues from the manufacture of what is known as 2,4,5-trichlorophenol. My curiosity was keen because I knew that this substance, which Hooker produced for the manufacture of the antibacterial agent hexachlorophene and which was also used to make such defoliants as "Agent Orange," the herbicide employed in Vietnam, carried with it an unwanted byproduct technically known as 2,3,7,8-tetrachlorodibenzo-para-dioxin, or tetra dioxin. The dioxin potency of this isomer was nearly beyond imagination. Although its toxicological effects were not fully known, the few experts on the subject estimated that if three ounces of tetra dioxin were evenly distributed and subsequently ingested by a million people, or perhaps more, all of them would die. It compared in toxicity to the botulinus toxin. On skin contact, dioxin caused a disfigurement called chloracne, which began as pimples, lesions, and cysts, but more important, could lead to wrenching internal damage. Some scientists suspected that dioxin caused cancer, perhaps even "galloping" malignancies that occurred within a short time of contact. It was dioxin, one to eleven pounds of it, which was dispersed over Seveso, Italy, in 1976 after an explosion at a trichlorophenol plant: dead animals littered the streets, and more than three hundred acres of land were evacuated. It was because of the dioxin contaminant that the spraying of Agent Orange was banned in Vietnam in 1970, when the first effects of dioxin on human beings began to surface publicly, including its powerful teratogenic, or fetus-deforming, capabilities.

The ban on herbicidal warfare involving Agent Orange was sparked by articles that appeared in *The New Yorker* before the by-line of Thomas Whiteside. Whiteside was a friend of mine, and when it occurred to me that Hooker might have buried trichlorophenol at the Love Canal, I called him for an informed viewpoint. "It's an extremely serious situation if they find dioxin there," he said. "This is most serious. If they buried trichlorophenol, there are heavy odds, heavy odds, that dioxin, in whatever quantities, will be there too."

After our conversation, I called Hooker. Its sole spokesman, Bruce Davis, then executive vice-president, was by now speaking to the media, but obtaining information from the firm was not the easiest or the most pleasant of tasks. Questions often had to be submitted days before they could be answered, as they were circulated through the hands of legal advisers and sometimes sent on to Hooker's parent company, Occidental Petroleum, in Los Angeles. Also, it was well known that Davis did not like my reporting, and there had been rumors that he and other Hooker executives had demanded that I be replaced. For several months these reports persisted, leading to tense moments between myself and the publisher. None of this added to the rapport I had with the company. I had posed two questions concerning trichlorophenol: Were wastes from the process buried in the canal? If so, what were the quantities?

On November 8, before Hooker had answered my queries, I learned that trichlorophenol had indeed been found in liquids pumped from the remedial drain ditches. Still, no dioxin had yet been discovered, and some officials, ever wary of more emotionalism among the people, argued that because the compound was not soluble in water, there was little chance that it had migrated offsite. At the same time, officials at Newco claimed that if dioxin was there, it had probably been photolytically destroyed. Its half-life, they contended, was just a few short years. When my story on the trichlorophenol appeared, Newco's project supervisor, Stanley Zawadzki, became extremely annoyed with me because, he said, I was scaring his workers—those who were digging the trenches. He stated flatly that there was no dioxin there. At the same time, one state official became so angered that I had publicly raised the possibility of the presence of dioxin that he threw that day's newspaper against the wall and stormed down a hallway at 99th Street School.

I could not accept the assurances that dioxin would not be found. I knew from Whiteside that there had been no known case where waste from 2,4,5-trichlorophenol had not carried dioxin with it. I also knew that dioxin *could* have become soluble in the groundwater and migrated into the neighborhood after mixing with solvents such as benzene. Moreover, because it had been buried, sunlight could not have broken it down.

On Friday, November 10, I called Hooker again to urge them to answer my questions. Davis came to the phone and, in a controlled tone, gave me the key answer: his firm had indeed buried trichlorophenol in the canal—200 tons of it.

I called Whiteside immediately. His voice took on an urgent tone as he said, "I consider this to be an alarming situation. Serious indeed." According to his calculations, if 200 tons of trichlorophenol were there, in all likelihood it was accompanied by 130 pounds of tetra dioxin, an amount that would equal the estimated total content of dioxin in the thousands of tons of Agent Orange rained upon the Vietnamese jungles. The seriousness of the crisis had measurably deepened, for now Love Canal was not only a dump for highly dangerous solvents and pesticides but a broken container for the most toxic substance ever synthesized by man.

I reckoned that the main danger was to those working on the remedial project, digging in the trenches. From the literature on dioxin, it was apparent that even in quantities too small to detect, the substance possessed vicious characteristics. In one case, workers in a trichlorophenol plant had developed chloracne although the substance could not be traced on the equipment with which they worked. The mere tracking of minuscule amounts of dioxin on a pedestrian's shoes in Seveso was of major concern, and according to Whiteside, a plant in Amsterdam found to be contaminated with dioxin had been "dismantled, brick by brick, and the material embedded in concrete, loaded, at a specially constructed dock, on ships, and dumped at sea, in deep water near the Azores." Workers in trichlorophenol plants had died of cancer or severe liver damage, and had suffered emotional and sexual disturbances.

Less than a month later, on the evening of December 9, I received a call from Dr. Axelrod. He asked what my schedule was like.

"I'm going on vacation," I informed him. "Starting today."

"You might want to delay that a little while," he replied. "We're going to have something big next week."

That confused me. "What do you mean by that?"

There was a pause on the line, then he said, "We found it. The dioxin, in a drainage trench behind 97th Street. It was in the part-per-trillion range."

News that dioxin had not only been identified but had actually migrated from the channel caused an immediate uproar among the residents, who were already picketing the construction site in an effort to force more evacuations. The newswires began to release a stream of urgent reports. On Monday at 5 A.M., in wet sub-zero cold, several dozen residents arrived at the main gate to the remedial construction, carrying placards about dioxin and demanding that the state stop the trenching. They feared, with reason, that the tires of the construction trucks would spread dioxin around the outer neighborhood. By 8 A.M., discouraged that police would not allow them to stop the trucks, Marie Pozniak and Charles Bryan, a 100th Street resident, were arrested when they refused to leave the middle of the road. "My child's crying all the time," Mrs. Pozniak told reporters. "She wants to get out." During the next hour, four more people were arrested and charged with disorderly conduct, among them Patricia Pino. Then on Tuesday Mrs. Gibbs, pushing a baby carriage to symbolize birth defects, was herself taken to jail with seven other protesters. Before the week's end, seventeen had been arrested.

The state remained firm in its plans to continue the construction, and despite the ominous new findings, no further evacuations were announced. Work continued. And during the next several weeks, small incidents of vandalism occurred along 97th and 99th streets. Tacks were spread on the road, causing numerous flat tires on the trucks. Signs of protest were posted in the school. Meetings of the Love Canal Homeowners Association became more vociferous than ever. Christmas was near, and in the association's office at 99th Street School, a holiday tree was decorated with bulbs that said DIOXIN.

By the time of the first anniversary of the Love Canal evacuation, the residents still living near the abandoned streets had reached the breaking point. As the remedial trenching continued, strong fumes were released into the air, causing headaches, nausea, and fainting. A United Way office established in the vicinity to aid the victims was abandoned when some of its workers became ill. A 102nd Street resident, Ann Hillis, stomped into a state office near the canal and threw a book at an official, thereby venting her anger at the state's refusal to relocate her family despite evidence of a 30 to 45 percent rate of miscarriage in some sections of the neighborhood. She claimed that her first child had been born so badly decomposed that the doctors could not determine its sex, and that she herself suffered from a bladder infection. Finally Mrs. Hillis and several other families were temporarily moved into hotels by the state until they "felt better."

Health Commissioner Axelrod had complained of his dilemma: "But under what authority do we evacuate everyone? Do we evacuate all of Harlem because it has an infant mortality rate four times the state average?"

Many residents who had scoffed at the risks now began to fear for their safety. The state had announced new findings of dioxin in a construction tank at the southern end of the area, and then, at the end of August 1979, the school board closed another building, the 93rd Street School, located about five blocks from the northernmost part of the canal. Clearly, the perceived zone of contamination was rapidly widening. Many of those who lived near Black Creek, in homes larger and more luxurious than those nearer the canal, were growing fearful because of the odors they noticed in the air, and an increasing number of residents began to filter into the hotels under the temporary provisions. But by Labor Day of 1979 the state had forced them out of the hotels to accommodate the season's last rush of tourists. A group of 270 persons who refused to go back to their homes were accepted for several days at a Catholic school dormitory. The school year was beginning, however, and after a few short days the nomadic group was again without a place to stay and still adamant about not returning to the Love Canal. As the state continued to deny them permanent relocation, their mood turned increasingly hostile. Losing control at one meeting, Ann Hillis threw table knives and a can of soda at a state representative and, along with several others, had to be treated at a hospital for shock. A man rushed to punch the same state official. Another state employee was assaulted by a woman who had recently lost her fourth fetus.

Under this tension, the state had no choice but to allow the people to remain in the hotels until the construction work was finished. But in order to remain, they were required to have slips from their doctors saying that to return to their homes would jeopardize their well-being. Only with great reluctance, and after an intensive effort by a task force composed of fifteen churches and the Love Canal Homeowners Association, were physicians persuaded to issue such documents. By autumn about 110 families, at a cost to the state of up to $8,000 a day, were living night-to-night in local hotels which had lacked for business during the summer months but now, ironically, were faring well. Meanwhile the residents' lawyers prepared court action to seek permanent relocation and to sue the school board, the county, the city, the state, and the Hooker Chemical Company for more than $10 billion.

There were protests to follow, and of course there had been protests before. The Love Canal people chanted and cursed at meetings with state officials, cried on the telephone, burned an effigy of the health commissioner, traveled to Albany with a dummy of a child's coffin, threatened to hold officials hostage, sent letters and telegrams to the White House, held days of mourning and nights of prayer. On Mother's Day, 1979, they marched down the industrial corridor carrying signs denouncing Hooker, which had not issued so much as a statement of remorse. The federal government was clearly not planning to come to their rescue, and the state felt it had already done more than its share. City Hall was silent—and remains so today. Some residents still hoped that, by a miracle, a government agency would move them. All of them watched with anxiety as each newborn arrived in the neighborhood, and they looked at their own bodies for signs of cancer.

While the state health department had earlier done an admirable job of exposing the danger, it was charged that this was no longer the case once the state decided to purchase no more neighborhood homes. There may have been a number of reasons for this new posture, including the technical difficulties involved in such a complex task, but at certain junctures there also seemed to be a strong tendency to soft-pedal inconvenient facts that would increase the residents' alarm, or to release data late on Friday afternoons when those reporters most familiar with the situation were not prepared for the news. In the meantime, local physicians were loath to enter the matter, and Mrs. Gibbs charged the state with deliberately shielding the results of air and blood tests. (She also claimed that scientists friendly to her cause had been demoted or transferred, and that a sample of a fetal placenta had been mysteriously destroyed.) Dr. Paigen claimed that she had been harassed as a consequence of her statements on behalf of the homeowners. (Dr. Axelrod described these allegations as "grossly untrue.")

Mrs. Gibbs, with growing support from the neighborhood, waged a relentless battle to discover the full extent of the contamination. Thoroughly familiar with the nuances of the situation, both on a scientific and a political level, she proved to be a remarkable adversary to the state and to Hooker, and as the homeowners became increasingly conversant with the effects of toxic chemicals, they grew more adamant in their refusal to be left behind. By this time dioxin had been found in Black Creek, where the Kenny boy had played, and some scientists said the area had a high potential for cancer. The residents held motorcades through the city. Mrs. Gibbs traveled to Washington to espouse their cause. And once in hotels under the temporary provisions, more and more people vowed they would not return home. Finally Governor Carey, reacting to the daily pressure, agreed that the state would move 200 to

550 additional families during the next two years. And so, today, the 1,000 or so outer-ring people are in line for a relocation by the state while others even farther away have begun to be concerned for *their* safety. The victory was a muted one. For those who have been evacuated will never feel confident that their severe physical and psychological scars will ever heal.

Timothy and Karen Schroeder now live on a twenty-five-acre farm thirty minutes to the northeast of Niagara Falls, in a land of alfalfa pastures, tall summer corn, and clear creeks. They have two pigs, Hambone and Simone, that they plan to butcher for their meat. "And twenty-two chickens and three calves," Karen told me. For entertainment there is a pony named Dusty. For water there is a private well.

On a trip to the farm, where they moved after the 1978 evacuation, I found Karen, a stocky, ruggedly handsome woman with a dark complexion, walking barefoot in the large meadows and catching toads for her youngest son, Timmy. "Now I step in cow crap," she said jokingly. "Not chemicals."

Karen pointed to a neighbor's yard, where her daughter Sheri was riding a swing. The girl enjoys roaming the acreage, and can do so with less fear of being hit by a car she cannot hear approaching. In a special class for the handicapped, Sheri has been winning awards for her skills in reading and writing. Another daughter has developed rheumatoid arthritis, but over all the family's health has greatly improved. There is far less coughing and not as many headaches nor as much fatigue.

Not far from the Schroeders' new homestead is a reservation assigned to the Tuscarora Indians, a people who long ago were pushed back from the Niagara River when settlers saw energy in the fast waters. One of the Indians, a thirty-year-old artist named Richard Hill, lent to a reporter his perspective on the Love Canal and why it happened. "At the first Thanksgiving we taught you to be thankful for what's here, to cultivate it and not abuse it. It never was our belief to treat Mother Earth that way. The earth is our last home. Now we see the results of technology gotten out of hand. And now, the earth is reacting very badly."

3

BLOODY RUN

There were many people—in government, in industry, in the media, and in the general community—who felt that the destruction of a small community by the Hooker Chemical Company's waste was a tragic but unforeseeable accident, one that could have resulted from almost any large plant's operations in the days before the environment became an issue and when the potential dangers of ground pollution were all but unknown. Not corporate insensitivity and ruthlessness but ignorance and uncontrollable circumstances seemed responsible for the Love Canal victims.

However, it soon became apparent that Love Canal was in fact part of a general Hooker pattern. In the year following the crisis at 99th Street, the citizens of Niagara County were to find that, in a manner similarly disregardful of their health and welfare, Hooker had dumped massive amounts of harmful residues in several other parts of town, destroying the subterranean water, deeply scarring waterways and soil, and endangering many more human lives.

Hooker's history went way back: established in the first decade of the twentieth century by Elon Huntington Hooker, it was one of the first companies to capitalize on the flow of electrical power. Hooker set up shop in a three-room farmhouse and called his company the Development and Funding Company. The first plant began operation in 1906, producing caustic soda from salt brine. Hooker's funds came from a wealthy group of investors who, for collateral, had settled for nothing more than Hooker's "character and ability." Growth of the company was spurred by the firm's investment in what was called the "Townsend cell," a device which electrically set off chemical reactions that separated the components of salt brine, leading to the production of the all-important chlorine; by 1978 the company was manufacturing chlorine at a rate of 816,928 tons a year. In 1909, the Hooker Electrochemical Company was chartered as a subsidiary of the Development and Funding Company, and by 1915, with a capacity at least four times that of the original plant, it built the nation's first plant to extract, from coal tar, substances vital to the production of war materials. Public trading of its stock commenced in 1940 after annual sales had reached $20 million.

As its income continued to leap upward, the company expanded to other parts of the country—Tacoma, Washington, and Montague, Michigan—until many years later it would operate not only in the United States but in Canada, Mexico, South America, Europe, and the Middle East. During the 1960s, after it was purchased by the Occidental Petroleum Corporation (whose chairman was the well-known Armand Hammer), the company was renamed the Hooker Chemicals and Plastics Corporation. Processing more than one hundred products in its mainland facilities, the company placed its copious supplies of chlorine at the service of a modern alchemy that yielded a bewildering variety of pesticides, which were snapped up by farmers and gardeners intent on exterminating every bug. Despite indications that many of its chemicals were highly toxic, perhaps even carcinogenic, Hooker forged ahead in its operations. Restructured later on as the Hooker Chemical Company, with many groups under its name, it amassed net sales of $1.7 billion in 1978, with a net income of $38.5 million. Naturally there were enormous quantities of wastes from such an operation, and the goal was to get rid of them cheaply.

Downriver from the falls, completely crosstown from William Love's abused canal and overlooking the steep gorge of the Lower Niagara, flowed Bloody Run Creek, a small tributary which nearly reached the proportions of a minor river as the snow melted but which dwindled once spring turned warm and dry. On September 14, 1763, two hundred Seneca Indians engaged in battle with twenty-five English settlers not far from the creek, and all but two of the white men were killed and thrown into the river ravine. Now, two centuries later, the white man, using the resources of science, was to supply a second justification for the creek's lurid name. Beginning in the 1960s, the stream assumed hues of orange and red at various intervals: chemicals were seeping from a second Hooker dumpsite. This waste pile would eventually be considered a more serious long-term problem for the environment than even Love Canal. It was thus the second in a long chain of horror stories now emerging about Hooker.

Near Bloody Run were several small factories, a private university, and a residential area. Most of the homes were across a large thoroughfare from the creek, but in its immediate vicinity there was a cluster of about fifteen houses, set among fields of goldenrod and stands of cattails sprung from several swales. Nearest the creek was a large corn patch and a small dwelling shaped like a barn, surrounded by a random array of firewood, bricks, and children's toys. Inside, it was equally cluttered. Stacks of personal records, dishes, and medicines filled the counter in the kitchen. The living room was also stuffed to capacity, with torn chairs and a rack of well-worn coats. There were four small bedrooms in the dwelling, and ten people lived there, in conditions less than lavish.

These ten people were Fred Armagost, a short, tense man who dressed in the attire of a farm worker, and his children and grandchildren. Half of his family had breathing difficulties; they wheezed as if affected with asthma or the tightness of allergic bronchitis. When the wind blew in the direction of the house and the air was moist and warm, noxious odors from the creek spread over the small area, resembling closely the fumes of the Love Canal. The family's pediatrician, Dr. James A. Dunlop, thought it likely that these vapors were contributing to if not actually causing the ailments, for neither he nor another doctor who examined the children could explain the high incidence "on any other than extrinsic causes." One of the grandchildren had developed a persistent angry rash of purplish blotches and open lesions on her body, and there was a history of kidney disorders and miscarriages in the family, all of which Fred blamed on the odors of the creek and nearby dumpsite. To declare his beliefs, he had erected on the front fence a hand-painted sign that read, "20th Century Massacre at Bloody Run—Product by Hooker."

"In 1963 or 1964 I called the state and was told there was nothing to worry about," said Fred bitterly. "There was nothing to worry about the landfill, they said. In 1973, or maybe a little earlier, a guy came over here from the regional office [of the state Department of Environmental Conservation] and right at the end of the culvert he got down on his knees and took a drink of that water. He said, 'That's pure. There's nothing in there.' This is when I really got pissed off. Why, we had a cat, and one day it comes back with this tarry crud on it. It lost its fur and its teeth. Five days later it died. We found it on the porch. Then there was the three goats I had that died. The one I know got down in that stuff and that sucker died in three days."

Certainly it had not been wise of the government inspector to drink water from Bloody Run Creek. Fred showed me to the murky waterway and, with an oak branch, stirred up the muck under the shallow water. A black, oily substance rose and spread across the surface, unfolding into a bright violet sheen with flecks of blue and brown. As he continued to agitate the sediment, I moved downstream to collect a quart of the water in a glass jar. We returned to the house, and Fred brought from a storage shack another jar filled with blackened mud he had previously scooped from the creek to show local authorities, who were not at all interested at the time in his dilemma, including the fact that Bloody Run often overflowed onto his property, staining his vegetable garden.

On July 1, 1978, I spoke with Steve Odojewski, a chemist at the private laboratory where the mud and water from Bloody Run were being analyzed. Odojewski was extremely concerned about what I had brought him. "This is very serious stuff," he said. "The concentrations indicate a serious problem. If these things were coming through an industrial discharge pipe, they'd seal it off. I couldn't have expected this quantity and variety." The analysis, done with a gas chromatograph–mass spectrometer, showed part-per-million amounts of lindane, C-56, and the pesticide Mirex, as well as other dangerous chlorinated hydrocarbons. Odojewski advised me not to put my hand in the creek again.

Most significant was the detection of Mirex. For years,

the Department of Environmental Conservation in Albany had been trying to locate the source of Mirex contamination of fish and sediment in Lake Ontario near the mouth of the Niagara River. The future of the huge freshwater lake, with its more than 7,000 square miles of surface and the 5 million people along its shores, was clearly at stake. The reason was the extreme toxicity and persistence of this compound, which, before it was restricted, had been employed in the South to control fire ants and its active ingredient also used as a flame retardant and plasticizer. A white powder in appearance, it was chemically constituted when two molecules of C-56 were combined in the presence of an aluminum chloride catalyst, and was so similar to another pesticide, Kepone, that only the greatest analytical precision could distinguish the two.

Fortunately, no outbreak of Mirex poisoning had been known to occur in humans, so that evidence of its toxicity to the highest primates was scant. But the effects of Kepone were all too well known. In Hopewell, Virginia, at a plant that manufactured Kepone from materials sent by Hooker among other sources, approximately 60 of the 149 employees who came in contact with the substance during a sixteen-month period were hospitalized with nerve damage, loss of memory, slurred speech, and severe weight loss, and there was also a high incidence of sterility and liver damage, enough to draw Hooker into a massive lawsuit that it settled out of court with the workers. While Mirex did not have so blatant a record, its harmful effects on laboratory animals were well documented. In female rats, a dosage of 12.5 milligrams per kilogram on the sixth through the fifteenth days of gestation caused maternal toxicity, pregnancy failure, decreased fetal survival, reduced fetal weight, and an increased incidence of other anomalies. There was also what was considered substantial evidence that Mirex is a cancer-causing agent in humans. Unlike most chemical compounds, Mirex continues to accumulate at higher and higher levels in the brains, muscles, and skin of mammals fed constant increments of it in experimental diets, without reaching a saturation plateau. It is highly resistant to chemical and metabolic attack, remaining in animals and the environment for decades with a tenacity greater than that of other chlorinated hydrocarbons such as DDT. Because of these factors, the EPA and United States Food and Drug Administration had set at .1 part per million the value level above which fish with Mirex accumulated in their flesh could not be sold interstate—a figure fifty times more strict than the acceptable limits for PCBs and DDT.

When Mirex was discovered in Lake Ontario bass, perch, and trout at levels far in excess of the limit, a great alarm was rung in Albany, and on September 14, 1976, DEC Commissioner Peter Berle placed a temporary ban on fishing and issued an order that prohibited possession of certain species. Though they had found some probable outlets, including discharge points at the Hooker plant itself, officials concluded that there must be origins other than those which had been positively identified. Now, after the Bloody Run tests, it appeared that another major source —one with the potential of continual contamination—had been located.

Bloody Run Creek originates in the area of the landfill. The dumpsite itself, referred to by Hooker as the Hyde Park "landfill," is essentially a seventeen-acre triangular plateau rising fifteen feet above the surrounding terrain and surrounded at the time by a wobbly fence and deteriorating berms and dikes. It was begun by Hooker immediately after the Love Canal filled up. Its height, allegedly in violation of local ordinances (it had provoked a $25 million lawsuit by the town of Niagara in 1979), meant that there were drums at once below and above the surface. When precipitation was heavy, large amounts of contaminated runoff poured from culverts leading from the landfill into a drainage basin on the north side, then on to Bloody Run and the Lower Niagara River. The gravel around the dump was tainted orange. Over the property near the landfill there hung the distinctive smell of C-56. According to one resident, dogs had been known to burn the pads of their feet passing alongside the waste site, yet though the property was open to pedestrians, there was no sign warning of any dangers.

In the midst of the initial Love Canal publicity, and at the insistence of the DEC, Hooker announced that it was initiating remedial measures, at a cost of between $850,000 and $1.5 million, to cap the site with more clay and improve drainage methods. But this was only a temporary solution. The most grating fear was that, because of the area's geology and the fact that the landfill occasionally hit bedrock, the chemicals might have seeped through the sand and silt and clay, cut through the underlying dolomite with their acid edge, and contaminated the underground aquifer. Hundreds of thousands of people drew their drinking water from Lake Ontario, and several families had wells in the immediate vicinity of the landfill.

When I called the regional DEC office to report my findings, the officials seemed surprised, as if unaware of any contamination at Bloody Run. My impression was a false one. They had known of the pollution since 1976, when tests made by the Environmental Protection Agency indicated the presence of Mirex in a drainage basin across the street from the dump and at the head of the creek. Tests through 1977 had shown the presence of high concentrations of PCBs and tetrachlorobenzene, thousands of parts per million, in the creek. The DEC and the Niagara County Health Department, which had warned Hooker of violations at the site—a concern that arose because, in departmental language, it was "operated rather carelessly"—negotiated with the company for proper closure of the landfill.

But Hooker was "initially very uncooperative," said the state documents, and appeared to be stalling for time. The dumping ground was a convenient one, only a few miles from the plant, and simply unloading waste into the elevated pit was a cheap way to dispose of residues. Hooker persisted until at least 1975.

Approached with the Mirex findings, Hooker public affairs chief Charles Y. Cain denied knowledge that the substance had been buried near the creek. Two months later, faced with further evidence, the firm's executive vice-president and new public spokesman, Bruce Davis, finally admitted that Hooker had dumped Mirex. Only later did state investigators learn that 1,100 tons of C-56 wastes and 4,500 tons of its derivatives (among which is Mirex) were buried there, in both liquid and solid form.

Mirex and its related compounds were not the only cause for concern at Bloody Run. By Hooker's own figures, which must be considered conservative ones, there are 80,000 tons of various highly toxic chemicals in the Hyde Park fill—mercury and benzenes and various chlorinated compounds. Upon reviewing state tallies, I noted with alarm that 3,300 tons of trichlorophenol, the precursor of dioxin, were also contained in the landfill, an amount sixteen times that in the Love Canal and perhaps more than is known to be landfilled in one place anywhere else in the world. (Another contender for the dioxin crown is a 93-acre landfill near Little Rock, Arkansas, that was used to store wastes from the production of

2,4,5-T; here a stream had to be closed to swimming and fishing.) Breaking down the figures, Thomas Whiteside, the dioxin expert who wrote for *The New Yorker*, estimated that 2,000 pounds of dioxin had been buried near Bloody Run—an amount that, if administered directly, would be sufficient to kill more than 95 *billion* laboratory guinea pigs. Perhaps some dioxin had seeped into surrounding swales or, combining with solvents, had volatized into the atmosphere and rained down upon the land; this might be why so few rabbits were seen in the area.

Not long after I learned about these frightening statistics, I received a phone call from Joan Gipp, a councilman in the town of Lewiston, New York, who had worked closely with me during 1977 in exposing the practices of a large landfill operator in her township. She had secured a copy of a letter dated December 6, 1978, from Hooker to the EPA's document control officer in Washington. The second paragraph contained some fascinating information: "A spot sample was taken of leachate from the landfill lagoon and of sediment from a drainage basin located near the site which receives rainfall runoff from the surrounding area including the landfill and other industrial properties. The presence of tetrachlorodibenzo-para-dioxin (TCDD) has been detected. . . ." Tetra dioxin was leaching out of the landfill—and toward Bloody Run Creek.

By early 1979, the state health laboratory, employing the services of a special laboratory equipped to detect dioxin, announced that relatively high amounts of dioxin were present in the creek near Fred Armagost's garden—in the same vicinity where the state inspector had supposedly cupped his hands and tasted the water, deeming it pure. It was also discovered at the point where the creek cascaded over the gorge and into the river. A health hazard was declared immediately, and, in an unprecedented action, the state commissioner ordered that a fence be constructed around what once had been a natural stream, to keep both humans and animals away from the sediment, which Hooker had turned into as toxic a sludge heap as could be conceived by man.

As news seeped out about this second major Hooker problem, those who worked in the three plants adjacent to the landfill, Greif Brothers Corporation, Niagara Steel Finishing Company, and NL Industries, grew nervous that the vapors they had noticed drifting in from the spoils heap might have affected their health, too. All three plants were much closer to the dumpsite than any of the homes, flanking its immediate northern, southern, and western edges.

The first voices of concern were raised by two Greif Brothers employees, Lawrence Maillet, a safety officer, and Dennis Virtuoso, at the time president of the plant's union. Their apprehensions increased after dioxin was detected in a drainage basin that flowed alongside and under the plant. In the past, strong fumes had emanated from a manhold inside the Greif facility. When a plastic sheet was placed over the sewer opening, it had been eaten away by chemicals and was replaced with a steel plate. Few complaints had been made against Hooker, however; the plant was in the business of manufacturing metal drums, and its main client was Hooker.

Despite pressure from management and the older employees, who feared the plant would be closed and their pensions lost, Maillet and Virtuoso proceeded to take an informal survey of their co-workers' health. They noted what appeared to be a high incidence of disease among those who labored near the manhole or where fumes were detectable in the air. Skin rashes seemed to be prevalent;

among the 90 or so employees, there were at least 17 such cases just at the time of their review. There had also been 8 cancer cases in eight years. When that figure was later verified by federal investigators, they described it as "unexpectedly high." Maillet and Virtuoso, both in their mid-twenties and fearing for their young families, loudly demanded that government check the employees for other ailments, a plea supported by their union, United Steelworkers Local 12256. In 1979, after much struggle with reluctant authorities, the National Institute of Occupational Safety and Health arrived.

The first interim report filed by NIOSH on May 7, 1979, gave strong hints of a contamination problem. An initial survey of 30 of NL's 261 employees revealed 10 cases of skin rash, 6 cases of high blood pressure, 5 of respiratory ailments, and cases of other dysfunctions reminiscent of the Love Canal maladies. Meanwhile, the union at NL had found that out of 91 respondents to its own survey, there were 45 cases of sinus problems, 41 of skin rashes, and 14 of cyst eruptions. That the plant's rank and file should suspect the landfill as the culprit was only logical: while all three plants used chemicals—barium, lead, methanol—that could also have been a source of problems, there were many times when landfill fumes gagged the men to the point where they walked off the job. "Every day, there was that smell," a former union vice-president told me. "It was nauseating, burnt your eyes. When I told management, they said, 'You can't challenge Hooker.' " So upsetting became the situation that on August 15, 1972, NL's plant manager, Dom Laurie, sent a letter to Hooker stating that the dump presented a "critical" and "extremely dangerous condition for our plant and employees." Daytime workers were leaving the facility, and the night crew had spent their time coughing and nursing sore throats. Contaminated runoff from the poorly maintained site had seeped into the plant itself, causing floor tiles to peel off and ruining paint. Laurie urged that the dumping be stopped; nevertheless, Hooker continued it for several more years.

At Niagara Steel Finishing, which had a back fence that physically abutted the landfill close to a collection pond where dioxin had been detected, the Oil, Chemical and Atomic Workers Local 8-778 brought in its own physician, Dr. Christine Oliver of Boston, for a review of worker symptoms. At a local hospital, she directed a medical evaluation of 42 workers, taking X-rays, liver and kidney function tests, platelet counts, and urinalyses. The results hardly varied from those at the other plants; 21 respiratory cases, the same number of arthralgia cases, and 20 cases of skin disorders were reported. Headaches and limb numbness were commonplace, and there were indications of abnormal liver function. Quite unexpectedly, Dr. Oliver found a history of urinary tract infections in nearly 24 percent of the men. When she compared these findings with those from a group of 379 outpatients at a Boston hospital who had been described as "rather sick," she found the factory workers near Bloody Run to have more problems.

Dr. Oliver concluded:

> It is possible, then, that the findings of the nasal mucosa and septum, of the respiratory tract, and in some cases, of the skin, are the result of welding fumes [inside the plant]. That the abnormalities were not limited to welders, however, suggests extensive exposure to welding fumes throughout the plant—including the office—or an alternative explanation of the findings. Likewise, individual exposures to ethanol, and medication and/or

> previous history of hepatitis could explain the abnormal liver findings. But that these individual exposures have coincidentally produced the high rate of abnormalities observed for the group is unlikely. The abnormalities in liver function observed in residents of the Love Canal area in Niagara Falls is [sic] significant in this regard. It seems likely, then, that indirect occupational exposures have produced many, if not all, of the positive findings (with the exception of decreased hearing and "metal fume fever"). Specifically, these indirect occupational exposures are to chemical residues in the Hyde Park landfill.

But Hooker played no favorites, not even sparing its own workers. Some of its waste, which was not spread around to endanger the surrounding communities, was deposited on its own back property with the usual lack of precautionary measures. One of its largest depositories was located approximately three miles upriver from the falls and within 200 feet of the river's edge. By Hooker's code, this general dumping property was known as the "S" and "N" plant areas. There, workmen dumped and poured more than 70,000 tons of materials (containing chlorobenzenes, C-56, and trichlorophenol) directly into the slag, stone, cinders, gravel, and abrasives that constituted the land, permitting rapid chemical seepage. The dumps had been closed when the state began construction of a highway nearby that would have allowed travelers too plain a view of the grimy conditions. The surface was then covered with fly ash and cinders, which readily absorbed the rain. When the weather was cold, reactive steam could be seen rising from the surface.

Because of their proximity to the river and exceptionally poor construction, the "S" and "N" areas were serious environmental problems. The ground was saturated with water and subject to the caprices of a current nearby. But even worse, and unknown to the populace, the waste heaps were located across a small road from the drinking-water treatment plant that served Niagara Falls and some of its suburbs. In short, Hooker had buried massive amounts of the most toxic chemicals in frighteningly close proximity to the water supply serving 100,000 people. And Hooker had done all this knowingly from 1947 to 1975.

In August 1978, I heard a rumor that a scuba diver sent into the water plant's shoreshaft to report on its maintenance conditions had returned to the surface with a can filled with black sediment that had accumulated throughout the line and smelled much like Hooker's product line of specialty chemicals. There were significant amounts of it in the intake pipelines, and it reached up to the equipment that filtered the water.

With a feeling of urgency, I called City Manager Donald O'Hara and the public works director, Robert Matthews, to find out if the report was valid. Both assured me there were no major problems at the water plant. O'Hara said the diver had come back with "nothing but mud," while Matthews was somewhat more descriptive, saying that the sediment "stunk a little." They argued that it was merely the accumulation, for many years, of particulates from the atmosphere and riverway.

At 4 P.M. the same day, I received a call from my newspaper publisher requesting that I be in her office within thirty minutes. I arrived to find O'Hara, Matthews, a city biochemist, Hooker executive Bruce Davis, and a company public relations man present. In a very low-key and circumlocutionary fashion, they commenced to contradict what I had been told earlier. Yes, they said, the sediment from the intake was "slightly" contaminated with Hooker chemicals. In fact, they informed me, the city and Hooker had jointly tested the sludge and identified traces of Mirex, C-56, and other potentially toxic substances, including benzene derivatives. "But I'm satisfied the drinking water is totally safe," O'Hara stressed. "This is something that could unnecessarily scare everyone to death. It shouldn't."

Where did the contaminants come from?

The city was uncertain. And Hooker had no idea.

It was several days later that I learned about the "S" and "N" areas. When I approached the city about the dumping, they again refused to implicate Hooker, saying they doubted that the fills were leaking chemicals into the pipeline a short distance away. I was able to locate Elliott Lynch, a former chemist at the plant, who said that strange odors had pervaded the water plant beginning in the late 1960s and that a "black, tarrish sludge," which he described as "identical to what was over there at the dump," was found in the pipeline. "It [the "S" area] was leaching, it is leaching, and it will continue to leach until somebody does something about it," Lynch said. At the same time, I obtained an internal Hooker memorandum that showed that as early as 1976 a city official had suspected that Mirex was infiltrating the water plant property.

Matthews continued to dismiss the notion of an imminent hazard. He said the city might have to line its pipeline with impermeable material or construct a drainage system around the water plant as it had around the Love Canal, but he insisted it was not a threatening situation, remaining loath to blame the powerful chemical firm. In Washington, however, EPA officials feared that a sudden slug of contaminated groundwater would penetrate the intake lines, and they were weighing a declaration of "imminent hazard" for the site, while in Albany there was discussion of relocating the water plant. (Extensive testing programs were initiated to monitor the development of the contamination, and remedial measures such as lining the pipeline with metal were decided upon eventually as a temporary solution.) When months later Matthews was questioned before a congressional oversight committee investigating dumps, he was roundly attacked for "avoiding the obvious" and trying to "keep the heat off Hooker." Another issue was the city's practice of taking its samples to Hooker for analysis, which one congressman, Albert Gore, called "incredible," along with the other attitudes prevalent in Niagara Falls. Under intensive questioning, Matthews finally admitted it was "highly likely" that the dump was indeed leaking into the water plant.

Because city analyses of the drinking water showed extremely low levels of bacteria and supposedly no single chemical present in quantities exceeding danger levels, Matthews and others at the plant maintained their optimistic pretenses, describing the water as pure as any in the country. The statement may have been accurate with regard to bacteria content, but certainly not otherwise. In one period, tests showed traces of C-56 and trichlorophenol, and a level of trichlorobenzene higher than in any other of the nation's cities—in the treated water. No one knew what cumulative effect the dozen or so compounds, blending together and reacting, would have on the populace. Likewise, there was no way of knowing if perhaps the levels were far higher during lengthy periods of time when the water was not being actively monitored. The possibility remained of a temporary but massive infusion of waste chemicals. It was already a matter of record—test wells had been drilled near the shoreshaft—that contaminants had reached the water plant property, enough so that one monitoring well gushed with noxious liquids. So there was the distinct chance that the thousands of people supplied by the line had ingested, or would at some point be ingesting, in however small quantities, a concoction of compounds suitable for advanced chemical warfare.

The second week of September 1979, a full thirteen months after the problem was first discovered, Matthews finally announced that he was opening an emergency forty-eight-inch intake line away from the contaminants while remedial work proceeded on the main tunnel. The plan was to reduce water pressure in the gravel bed beneath the tunnel, creating an artificial pressure gradient that would prevent any more toxicants from entering the supply. There were no plans to compel Hooker to remove its chemical outlet. The mayor's only public words for Hooker were words of praise, and he proudly participated in the ceremonies initiating construction of the new Hooker headquarters downtown. When Love Canal residents appeared at the festivities to protest the company's attitude toward them, the new city manager, Harvey Albond, warned them not to cause a disturbance.

Nor was this the whole of the Hooker tale. There were other Hooker dumps along or near the water in Niagara County that were creating environmental distress. They too were inadequately sealed and had been maintained for years at minimal costs, despite the specter of human injury. Key among them was what was known as the 102nd Street dump, located just south of the Love Canal, on the banks of the Niagara River itself. It had been used by Hooker until 1970, when the United States Army Corps of Engineers, concerned that waste materials were finding their way to the river above the water intakes for the city, declared that it was being operated illegally and ordered it closed down. At one time Hooker tried to sell it to the city.

On March 28, 1966, the Hartford Insurance Group sent

a letter to Hooker's safety supervisor at the time, Carl Olson, urging that the firm place warning signs at the dumpsite and enclose the sprawling twenty-acre dump with a fence. Phosphorus was said to have been scattered on the surface, and children were playing there. Despite reports of small explosions at the dump, Hooker did not take the insurance company's advice until the summer of 1967. By that time, a most unfortunate accident already had occurred: thirteen-year-old Michael Gigliotti, walking across the dump with friends, was caught in a sudden barrage of chemical explosions set off by the mere friction of his feet. He was rushed to the hospital, where he remained for days, with severe burns on his legs and other parts of his body. Michael explained to me: "We went to the city docks to watch boats. We were exploring fields. All of a sudden, when we got to the river, there was a small explosion, a firecracker sound, under one of my friends' feet. Everybody ran, tore out of there. The way I ran, every step I took, there was an explosion. White smoke. My chest and face were burning from the chemicals. There must have been fifteen explosions. It got hotter and hotter, so I couldn't see, I couldn't see for three days after that. I was in the hospital for nearly three weeks, critical condition. The guy who drove me said the skin just rolled off my back in the back seat, on the way to the hospital."

In addition to its primitive waste "technology," Hooker also had created unhealthy conditions in the air of the city and in the workplace through the regular operation of its main facility, on Buffalo Avenue in Niagara Falls. And again, Hooker was fully aware of its responsibility. This was shown most vividly in April 1979 in a document, "Operation Bootstrap," that I had obtained from a former Hooker engineer. The document, an internal study aimed at outlining the plant's conditions in 1975, detailed what the authors described as "deplorable" working conditions and significant leaks into the city's general atmosphere of mercury, phosphorus-based gases, and pesticides. Those who had worked there told me of high rates of leukemia and skin eruptions they called "the Hooker bumps"—growths and lesions that afflicted those working near the trichlorophenol vessel and that resembled descriptions of dioxin-induced chloracne. The report's authors and the employees attributed the serious plant problems in part to broken-down equipment pushed beyond its capacity, demoralized foremen, and undertrained workers who knew nothing of the dangers presented by substances they regularly handled. "We used to test Mirex with our bare hands," one man told me. "We made snowballs out of it to see if it had crystallized." A young woman said that after working with benzene and suffering through a stillbirth, she asked for a transfer to a safer part of the plant, a request that met with management resistance.

The "Bootstrap" document was also significant to those who lived near Hooker's smokestacks and intertwining pipelines. Frequently through the years, Buffalo Avenue and the Robert Moses Scenic Parkway were forced to close temporarily because of large emissions of chlorine gas or other noxious fumes. The worst incident occurred on the evening of December 14, 1975, when a Hooker tank car containing process chlorine exploded into fragments, alternately reddening the night sky and causing an acetylene-like glow. Small rumblings like distant thunder were heard, followed by deafening blasts so powerful as to knock the shoes off one passing pedestrian. Propelled by a southwest wind, the yellow clouds of gas drifted for a mile across the city before settling onto a department store, felling clerks and customers alike, and causing terror among those it passed. More than 90 people were treated at the hospital after the explosion, and of these 4 were taken to the morgue.

Some other grave health problems existing in Niagara were hard to trace to particular causes. Between 1973 and 1975, the state's Cancer Control Bureau conducted a provisional tabulation of diagnosed male cancer cases in Niagara County that was both a well-kept secret and a potential cause of consternation. Cancers of the lung and bronchus during that short time were 33.7 percent above what would be expected; leukemia, 54.5 percent; kidney cancer, 44 percent; cancer of the pancreas, 27.1 percent; stomach cancer, 20.2 percent; and all types together, 23.6 percent. Later studies were to show a drop in some of the percentiles, and state officials were quick to discount their study's worth. Still, it appeared obvious something was affecting the county's cancer incidence, something widespread and unseen.

4

THE BARRELS OUT BACK

In the absence of effective legislation protecting the environment, Hooker had despoiled land in the United States far beyond the soils of western New York. Communities from Florida to California where the plants were located also suffered badly poisoned groundwater and a fouled biosphere. The pattern confirmed that systematic and arrogant attitude within the firm that condoned the degradation of the earth's green mantle and the violation of the nation's laws.

Hooker's disregard of public welfare was displayed nowhere more clearly than in a western Michigan town called Montague. The Hooker complex there was just north of White Lake, a renowned fishing channel, and consisted of freshly painted one-story and two-story buildings fronted by a luxuriously large lawn and a sign surrounded with healthy evergreens. The complex was set incongruously in a tourist area spotted with resorts that featured boating, swimming, and tennis, among the many moss-coated oaks that lined a winding roadway behind a shoulder of upturned sand. This was history-rich Old Channel Road, which had once been the site of an Ottawa Indian burial ground as well as of an eighteenth-century Potawatomi village of dome-shaped dwellings enclosed in a palisade. Later came the lumberjacks and millwrights, who chased the Indians away, and still later, in the 1950s, came Hooker.

On August 17, 1977, Warren Dobson, a former operator at the factory, broke the time-honored secrecy among plant employees and exposed the most acute environmental situation in all the state. Pricked by his conscience, Dobson charged Hooker with a number of unsafe practices in the production of the pesticidal intermediate C-56, a compound that, in gas form, has an established threshold exposure limit ten times more stringent than the standard set for phosgene nerve gas. Dobson claimed that C-56 had been allowed to seep through ceilings into work areas below, vent into the open air despite the plant's proximity to homes just across and down the road, and discharge into nearby White Lake, a great tourist attraction. Chlorine gas* was also allowed regularly to escape into the atmosphere, Dobson testified, adding that he was instructed by his superiors to "act as though the white vapor escaping from said pipe was steam" and to try to "convey the impression that he didn't know any toxic vapors had escaped" in the event that residents near White Lake complained of inhaling chemicals. When Dobson questioned these practices, he was told, "This is not a chocolate factory. . . . We got to make money." His complaints ignored, he quit the job after six years of service.

The air was indeed a problem. A sixty-nine-year-old

* To comprehend what massive emissions can mean to a community, one has only to read an account of the first full-scale gas attack of World War I. On April 22, 1915, in the Ypres salient on the French-Belgian border, the Germans released 170 tons of chlorine gas into a light wind along a four-mile front posted by Canadian, French, and British troops. The Allies counted 5,000 dead among 20,000 casualties; according to government literature, the faces of the dead were black and contorted in pain, their lips coated with foam and blood sent up by their burst lungs.

man who lived immediately west of the factory found the fumes so repulsive that he sometimes drove into downtown Montague to sleep in his car. He blamed the emissions for killing a large swatch of his trees, and he complained of extreme swelling of the stomach, an enlarged heart, and facial outbreaks so severe he was ashamed to attend his favorite social function, the fish fries at the local Veterans of Foreign Wars hall. His dogs had suffered throat disorders and cancer, he said; his current pet had a red growth on its tongue.

But to the Michigan Department of Natural Resources, aware of the hazards associated with C-56 and already disenchanted with Hooker because of other environmental transgressions, the sixth paragraph of Warren Dobson's affidavit was of more concern. It described a secret dumping ground in the back northern area of the plant's 880-acre tract, where 55-gallon drums of C-56 "still-heel" residues and other wastes had been slashed open with an axe and simply left on the surface, so as to allow the contents to drain into the sandy surficial layer of earth. From those barrels, deteriorated with rust, substances of frightening toxicity escaped, tarry mixtures resembling black molasses and emitting strong fumes. In testing at Michigan State University's Pesticide Research Center, these wastes were found to be highly toxic to fetuses and, according to the scientist studying them, as capable of causing cellular mutations as anything tested in the laboratory. In all, approximately thirty chemicals were being unleashed into the environment, and in staggering amounts.

Hooker knew as early as 1968 that it was contaminating the groundwater below its plant from sludge lagoons that were scattered around the property and contained salt. Yet when I visited Lansing, the state capital, two officials of the DNR, Andrew Hogarth and Dr. James Truchan, told me the company at first had claimed to have no knowledge of widespread underground contamination nor of the patterns of subterranean flow. They said Hooker had attempted to conceal its dumping even after Dobson's affidavit was made public, and that Duane Colpoys, then works manager at the plant, had assured them there were only "a few" or "maybe ten or twenty" residue drums in the back area—no problem, he argued. Not until the following March did DNR officials visit the site, and they were unprepared for what they found. Instead of "a few" barrels, there were an estimated 20,000 of them. Calculations showed that approximately 2 billion gallons of groundwater, spread under a 364-acre zone, had been poisoned by potentially carcinogenic compounds, at least a million gallons reaching White Lake each day.

On their way, the contaminants were infiltrating drinking wells, creating an immediate health hazard. Between Hooker and White Lake was a residential subdivision called Blueberry Ridge, and near Old Channel Road, a spacious house with a large fireplace and an expansive redwood deck from which the owner, W. G. Carroll, vice-president of a local cast iron foundry, could look out at the lake. "When we moved here we discovered the water was bad," he said. "The smell was terrible. It tasted awful. The guy who dug the well, he thought the same thing I did—we thought it was from the galvanized pipes. When it got hot, it got worse. You would take a shower and the steam . . . from the steam you could smell the fumes. It was nauseating. And it didn't get better. It got worse. So the guy who drilled the well said, Geez, it should have gone by now. I used to forget and make ice cubes with it, and you'd have guests over and put ice in a glass of straight whiskey and they would still say, 'What's in this ice?' "

The family drank the water only for the first several

weeks; soon they were forced to have water delivered and to use the faucets only to shower with and wash the clothes. At the time, Mr. Carroll said, Hooker was still popular in town, which made it difficult for him to complain. "We battled for a good five months about this," he said. "Colpoys' first thing when we finally reached him was, 'How do you know it's not DuPont (which owned a plant further up the road)?' Then he said maybe it was something in the ground itself. I told him, 'This is nothing Mother Nature made.' He denied there were any effluents or dumping. Then he came to the house and I gave him a glass of water. He wouldn't drink it."

In May 1978, the Muskegon County Health Department issued the following order to Mr. Carroll and his neighbors: "Until further notice, please be advised that the above contaminants [carbon tetrachloride and chloroform] are toxic chemicals that could have long-term effects on human health; therefore, you should not use your well for drinking or cooking purposes." Clearly, another source of water was essential. A report issued by the New York City engineering firm of Fred C. Hart Associates pointed out that if a 150-pound man drank daily a quart of water containing 25 parts per million of carbon tetrachloride, he would theoretically reach the "lethal dose low" within just four months. That level had been found in samples taken from the groundwater near Hooker. In another sample, 50 thousand parts per billion of tetrachloroethylene were found, quite in excess of the 35-part-per-billion standard the Environmental Protection Agency had declared to be the threshold above which aquatic life is threatened. Hooker paid to connect the residents to a municipal water line or to have water carted in.

To combat the various assaults on its manufacturing

techniques, Hooker chose two methods of counterattack, one public and one, apparently, quite behind the scenes. For its public response, the firm began a strong advertising campaign in the local newspapers. The advertisements had arresting headlines in tall, bold black print. IS IT TRUE WHAT THEY SAY ABOUT WHITE LAKE? one asked innocently. Another took a more injured tone: IT'S EASY TO HATE A BIG CHEMICAL COMPANY. In these public defenses, Hooker ventured that "toxicological evidence available to us indicates there is no hazard to fish or to people who eat fish from White Lake." It also said there was "far more chloroform in an ounce of toothpaste than in 80,000 gallons of lake water." And just to make sure its point of view was fully appreciated, it added that there are "150 different chemical substances in a potato. . . . One of them is arsenic, a deadly poison." One part per million of a chemical, it continued, was equal to only one drop of chocolate in 61 quarts of milk, while a part per billion was but one drop in 1,280 barrels of liquid. To cap it off, Duane Colpoys signed his name to a long letter admitting that "to say we don't have environmental problems at Montague would be misrepresenting the case," but at the same time pledging that "we are putting our house in order."

Hooker's other defensive measure, I was told, was less dignified than its newspaper pronouncements. According to a number of Montague area residents, employees had been known to harass or indirectly threaten those who dared protest the plant's environmental "problems," which had included violations of water-pollution control laws resulting in fines totaling more than $300,000. Carroll's attorney, Wint Dahlstrom, an ardent environmentalist who was involved in the Dobson affidavit, claimed: "There was a definite form of pressure. I lost a lot of Hooker clients. And there was direct hostility. I was at the White Sands Restaurant once, and Colpoys came over and insulted me in front of a group of people, and he gave me information that led me to believe he had collected a dossier on me. I used to tell people, 'Those bastards are lying to you.' " Dahlstrom also claimed that a friend of his, a partner in protest, was hissed at and kneed in the back during a state hearing involving Hooker. Later, said Dahlstrom, the man was clubbed over the head in his own backyard.

When the local weekly newspaper, the *White Lake Observer*, began to write about Hooker's pollutants, it too suffered intimidating assaults, according to the staff. The newspaper was housed in a small gray rectangular building on Ferry Street, on a ridge overlooking the stately pines that are a canopy for the lake's northern shore and the peaceful cottages. The cluttered newsroom on the first floor contained only five small desks, several telephones, and the tall, mustachioed publisher, Darwin Bennett. As I sat thumbing through a stack of back issues of his newspaper, my eyes caught an anomaly on the front page of the April 20, 1977 edition. There, at the lower left of the page, was a large blank space. This confused me until I turned inside and read a column Bennett had written to accompany the empty box. It began:

> This is a public apology to our readers for not printing all the news this week.
>
> We don't often make apologies like this.
>
> The role of any good newspaper, even a small weekly like ours, is to print all the news that affects its readers to the best of its ability.
>
> I'd like to feel it's like holding a mirror to the community in an attempt to reflect the true events that happened each week. It's a heavy responsibility; not everyone always likes what they see in that mirror.
>
> Recently, we've had the misfortune of witnessing a

large national firm going through the embarrassing ordeal of getting caught red-handed doing something against the law.

Its reaction may have been predictable—much the same reaction that a small boy might display upon being found out by his parents for a misdeed.

First, shocked innocence and denials of guilt, then the pointing of fingers at other parties it feels are responsible for its predicament, then anger and attempts to strike out at all who disagree with them, and finally, efforts at reprisal.

All to date have failed, but each has had its unsettling influence on the community.

Unfortunately, as a result, a great deal of heavy-handed pressure has been placed on many people in the community, and I believe the public should be made aware of it.

Bennett went on to mention threats of lawsuits, calls from outraged Hooker employees, petitions in support of the plant, and pressure applied to the *Observer*'s advertisers in an attempt, said the publisher, to discourage the paper from reporting on the firm's environmental misdeeds. He described those tactics as having "all the finesse of a playground bully, the public relations of Genghis Khan, and the techniques of Machiavelli." He ended his unusual complaint:

I'm ashamed to admit that this newspaper has been intimidated by all of this. We dare not oppose their wishes, for what chance would a small country weekly have against one of the nation's most powerful and wealthy industrial firms?

For weeks I've agonized over what to do about these attempts at censorship of the news.

As you'll notice from our front page, we will not carry any more news concerning state action against this company, just a blank story where the story should have appeared.

I'm sure the firm would rather you call them for any information regarding the state's latest action.

I sincerely regret being forced to take this action, but the underdog only wins in the movies, and my name is neither Woodward nor Bernstein.

"It was an insidious type of pressure," Bennett said, "where a pillar of the community would come up to me and say, 'Darwin, what are you doing?' Next, you would go into a store, one of your advertisers, and be told people supporting Hooker had come in with a petition and these people were told to 'tell Darwin.' But the advertisers didn't buy that. They weren't afraid of a consumer boycott by Hooker employees. Here we had only three articles and suddenly Colpoys had called and tried to put the pressure on. He said I didn't have my facts straight and I didn't know what I was talking about. I said I had a job to do and Colpoys said, 'We have a way of taking care of that, too.' "

The state was not as fazed as Darwin Bennett. Not long before my trip to Michigan, the DNR enforcement staff, in conjunction with the attorney general's office, had submitted to the Ingham County Circuit Court a strongly worded civil complaint against Hooker that requested civil penalties and damage payments for the dumping at Montague, along with the initiation, at company expense, of an extensive remedial-action program to purge the groundwater and isolate or haul away the contaminated earth. The cost to the firm would be between $10 million and $250 million. Michigan's legal actions were given muscle by a private company memorandum, dated November 4, 1955, and entitled "Study of Residue Disposal Systems." It stated: "The disposal of plant residues at the Montague plant is a major problem due to local and state ordinances"—an indication, to be sure, that the corporation's management was not unaware of activities at the plant. "This is the worst company I've ever worked with," protested Dr. James Truchan. "They lied, they seized a camera from one of our men, they obfuscated facts and hid information. I couldn't believe the dumping at Montague. It was incredible that they would dump that stuff on the ground. They argue that they were operating 'state-of-the-art' disposal. The heck they were."

In 1978, the Occidental Petroleum Corporation attempted, on unfriendly terms, to take control of a large Ohio paper company, the Mead Corporation, an endeavor that shed more light on its corporate practices. In the course of this battle, lawyers for Mead obtained a great number of internal Occidental documents pertaining to Hooker, many of them through Dr. Ed Kleppinger, a toxic waste expert from Washington, D.C. In those confidential papers were many examples of Hooker's callousness toward its environmental atrocities.

The same year the Occidental-Mead takeover was in the courtroom, a Hooker phosphorus plant in White Springs, Florida, pleaded no contest to charges of polluting the air with fluoride and, after misdemeanor criminal charges had been argued, paid a $38,000 fine, while the state appealed for another $8 million in civil penalties. As the documents obtained by Mead revealed, however, the Hooker plant had violated far more strictures than the agencies had suspected, including the emission of as much as 3,000 pounds of fluoride on a single day when only 34 pounds daily were permitted, and the discharge of fluoride and phosphates into perhaps the only unpolluted river remaining in the Southeastern flood plain, the famous Suwannee. Acid mists and sulfur dioxide also had been released into the air. Furthermore, the documents revealed that Hooker had done these acts to cut costs, at times with the full knowledge of the company's top echelon. In Houston, with the approval of two top officers, the company changed its phosphoric acid process to a cheaper one that released much more fluoride than was permitted. When it came time for inspections or for emissions tests, plant officers closed or altered the operations of spillways and smokestacks.

Hooker's inimitable brand of corporate citizenry extended to Lathrop, California, where a Hooker plant had dumped pesticides illegally into an unlined lagoon and discharged fertilizer and nitrates in such a way as to contaminate, in a three-mile radius, the plant's own wells and those used for cattle-watering on neighboring farms. In 1975, the papers showed, Houston management knew the Lathrop plant was violating California State Water Quality Control laws by discharging 5 tons of pesticides annually into the lagoon, from which they percolated into the ground, but allowed the practice to continue until 1976. The firm did not inform the state of the contamination until March 1979, when the state tracked dibromochloropropane, a pesticide banned in the state after allegedly causing sterility among workers at the same factory, in local wells. It was likewise seen that as recently as September 1978, an environmental engineer at the plant had pleaded with management to clean up the mess. But despite the firm's knowledge of unlawful conduct, and despite what should have been a lesson at Love Canal, funds for such a project were denied.

The memoranda demonstrated Hooker's full, chilling appreciation of the situation, as well as its public stonewalling. Robert Edson, the plant's chief environmental engineer, wrote on April 29, 1975:

> Recently published California State Water Quality Control laws clearly state that we can not percolate chemicals to groundwater. The laws are extremely stringent about pesticides. We percolate all of our gypsum water, our pesticide wastes, and 1 to 3 percent of our product to be found in the form of production losses. Not only must we stop this by law, but it will recover $20,000 to $40,000 a month in losses.
>
> Our neighbors are concerned about the quality of water from their wells. Recently water from our waste pond percolated into our neighbor's field. His dog got in it, licked himself and died. Our laboratory records indicate that we are slowly contaminating all wells in our area and two of our own wells are contaminted to the point of being toxic to animals or humans. THIS IS A TIMEBOMB THAT WE MUST DEFUSE.

Edson said he would not want to drink out of one neighbor's well, yet he did not mention the problem in his communications with the state water quality board. In a memorandum to the plant manager, James H. Lindley, he said he believed a report to the state was inaccurate because of the omission and felt uncomfortable signing it. "However," he continued, "I don't think it would be wise to explain the discrepancy to the state at this time." Edson further made clear, in a confidential 1976 interoffice memorandum, why it was that hazardous problems were not corrected. "Other companies' solutions are so expensive we haven't had anyone with enough nerve to even suggest that we follow their examples." In the same correspondence, he added, "Once again, this discharge point is less than 500 feet from our neighbors' drinking water well. I believe that we have fooled around long enough and already overpressed our luck." In 1978, another internal Occidental memorandum said that "in 1967 we told the state we would eliminate our discharge. We have not done so." In the face of Hooker's cover-up, the state of California finally sued the Occidental Chemical Company, the Hooker subsidiary there, for $45 million, charging that Occidental "knowingly and willfully withheld from the state the fact that it was allowing hazardous materials to percolate to the groundwater" near the plant.

Nor did the problems stop at Lathrop. In Louisiana, investigations were initiated into unproven reports that Hooker had dumped, without a permit, PCBs, asbestos, and other harmful substances at its Taft plant near the Mississippi. In Hicksville, New York, vinyl chloride from a Hooker plant was strongly suspected of contaminating still more wells. Interoffice memoranda from this facility also showed that the company was illegally dumping chemical wastes into the Oyster Bay town landfill in Bethpage—materials described at one juncture as "degenerative to bones in the hand."

Executives at Hooker continued to defend its practices as oversights or as "state-of-the-art" technology in line with the industry's standards. In an interview with *Chemical Week*, Donald Baeder, Hooker Chemical's president, said, "In our opinion, the company acted responsibly from beginning to end. There is not one incident where we didn't come forward to cooperate fully. Love Canal was the ignition for an environmental microscope of the company. Any responsible company would suffer under that kind of scrutiny." A former Dow Chemical executive, Zoltan Merzei, vice-chairman of Occidental Petroleum, said Hooker was "no more negligent than any other companies." To a limited extent, as we shall see, this was true.

In the same interview, Bruce Davis, who had so long defended the company's misdeeds in Niagara Falls, continued to insist that the firm was not responsible for the 3,000 or so people who had been, and may still be, in danger because of the Love Canal. His contention was that Hooker had dug five feet deeper into the clay of the canal and, once the drums were emplaced, covered the materials with the excavated clay, forming a "seal." The clay cover was disrupted, Davis charged, by local construction activities, and that was when the trouble began. Spent cake (residues from various processes) was found in backyards, he pointed out, indicating that the cover had been scraped off by private developers. Disposal at the Love Canal, Davis said, was "damn good practice even by today's standards." Only reluctantly did the firm grant its land to the school board, Davis claimed, and as for the 1958 incident, when it learned of children burned at the dumpsite, the firm thought that refilling had solved the problem for the time being. It seemed to Davis that illnesses reported at the canal were not as serious as many people thought, and that the houses could be used again. Speaking before a group of businessmen in Buffalo, Davis asserted that there was "no major health problem in Niagara Falls," adding that those residents who demanded evacuation had fears that were "based on ignorance" and "hysteria." "Hooker has been investigated, prosecuted, convicted, and sentenced in the media without having a chance to explain its side," he intoned, declining to mention the many times he himself had refused to answer a reporter's questions. During one two-month period I had attempted twelve times to obtain from Hooker any evidence backing its claim that there had been a clay cap on the Love Canal and that the firm had plainly warned the school board of the dangers. I was finally told that such documentation existed but that the time was not ripe for its release. In Niagara Falls, meanwhile, Hooker began an advertising campaign similar to what it had done in Montague, with bold black headlines more than an inch high proclaiming such things as "TRY TELLING BRUCE DAVIS THAT HOOKER DOESN'T CARE ABOUT NIAGARA FALLS." After detailing Hooker's financial importance to the community and its programs to correct environmental problems, this advertisement finished on a most ironic note: "When you get right down to it, you'd be hard pressed to find any group of people who care as much about the environmental and economic well-being of Niagara Falls as the people at Hooker."

Hooker continued to resist paying the high costs of permanently resolving its poisonous seepages. Instead of embarking upon an estimated $150 million project that would permanently remove the toxicants from Bloody Run, the "S" area, and the 102nd Street dump, the firm committed itself to only a fraction of that, spreading clay over the surfaces, upgrading drainage systems, and monitoring to see how far its dirt had traveled. It became obvious that there was no way of stopping such corporate assaults on the land without the public and its agencies carefully scrutinizing commercial operations and meting out to corporations, as to common criminals, the penalties they justly deserved. It would be difficult to determine the proper punishment for Hooker, for it could never be known just how many lives the company might have damaged and how many gallons of its exotic formulations would remain forever under the ground. But to forestall another firm's following its disastrous example, it seemed logical that Hooker should be fined the enormous amount it had saved by its inexpensive and unsafe practices. In the name of displaced and injured Niagara and Montague residents, Hooker officials who directly sanctioned illegal practices should be indicted on criminal charges.

Though no indictments were immediately forthcoming, the federal government did take an unprecedented if belated action through the Justice Department to compel Hooker to spend $117.5 million for a cleanup of four wasteyards in Niagara. Another $7 million was sought by the government for reimbursement of part of the evacuation costs at Love Canal, as well as company funding for a medical survey of the Love Canal and Bloody Run families for the rest of their lives. The Olin Corporation, which had wastes seeping from the 102nd Street dump, was also named in the suit, which was filed on December 20, 1979. According to the federal suit, these activities had created "an imminent and substantial endangerment to health and the environment."

Senator Daniel Patrick Moynihan, one of the very few politicians to be ruffled by the actions of Occidental's wayward subsidiary, expressed it well. "I hope Occidental pays every penny it owes to that community," he thundered. "There's no point in being restrained with this company. They don't understand it. It could end up costing a quarter of a billion, and if it does, Occidental Petroleum should end up paying for it. To dismiss it [Love Canal] as inconsequential verges on the unforgivable. They've taken enough out of that city, and left nothing but poison behind."

II

TOXIC AMERICA

5

IOWA: NOR ANY DROP TO DRINK

The gravest threat from hazardous-waste pollution seems not to arise from carcinogens soaking under back lawns or infiltrating cellar walls in odorous slurries. Instead, as in Montague, it is the effect of the dumpsites upon the subsurface flows of water. Groundwater, collecting between the particles of gravel and sand, or in pockets formed by porous rock, goes through underground channels that rise with hills and dip under valleys, visible only in the occasional presence of a marshland or spring. Like the grounds in a coffee pot, chemical landfills saturated with rain slowly drip their malign contents into those unseen aquifers beneath. And as a result, thousands of people throughout the United States, quite unknowingly, have been drinking water fouled by metals, pesticides, and potentially cancer-causing solvents used in industry to cut oil and grease.

When rainfall percolates through a landfill, it removes the soluble components from the waste, producing a grossly polluted liquid, the leachate. Depending on the dumpsite's contents, the liquid may contain viruses, bacteria, and decaying organic matter or, if industrial in nature, the various toxic brews. In humid areas, where precipitation exceeds evaporation, the amounts of discharge can be remarkable indeed. The Environmental Protection Agency's Office of Solid Waste has estimated that an average land disposal site, seventeen acres in size, with an annual infiltration of ten inches of water, can generate 4.6 million gallons of leachate a year, and can maintain this impressive productivity for fifty to one hundred years. A large landfill in New Castle County, Delaware, was found to be depositing 170,000 gallons of polluted water a day into an important aquifer, while in Islip, Long Island, a thirty-nine-year-old landfill had developed an underground plume that extended 1,300 feet wide, 170 feet deep, and 1 mile long, accounting for the destruction of 1 billion gallons of water.

Once it finds its way underground, leachate is nearly impossible to retrieve. The chemicals, shielded from the atmosphere, are not subject to the photolytic degradation of the sun, nor do they readily evaporate. Instead they cling tenaciously to the particles of soil, or remain in aquifers that stay in the same locations for great lengths of time. When the groundwater does move, it quietly filters into surface bodies of water, or finds its way into the kitchen sink.

Despite the importance of groundwater pollution—of greater consequence than more obvious problems, such as direct discharges into a river or refuse dumped into a creek—President Carter in 1978 produced an extensive water policy message without addressing the issue. In criticizing the omission, the United States comptroller general said, "The relationship between disposal practices and the effects on groundwater quality has generally been ignored. Land disposal sites for wastes are often located in areas considered to have little or no value for other uses; sufficient concern is not given for the type of soil on which they are situated or their proximity to water resources, particularly groundwater. Such improper siting, coupled with limited state enforcement of other standards and requirements, has resulted in groundwater contamination in some heavily populated areas throughout the country."

The comptroller general's office, apparently unlike the White House, was well aware of how important groundwater was to the populace. The office had done a natural survey and found that about 80 percent of municipal water systems, serving 30 percent of the actual population, depended entirely upon groundwater brought up from wells, with an additional 10 million families in more sparsely inhabited regions relying upon private wells. There were extensive reserves of groundwater in the nation, more than 180 billion acre-feet of it within a half-mile down, but much of it was not recoverable and it was unevenly distributed across the land. Moreover, its use was exponentially increasing each decade, creating shortages in the High Plains of Texas and dry portions farther west. In 1950 America used about 35 billion gallons a day; the figure is expected to triple by 1980.

While the supply of water dwindles, the spread of contaminants increases even more rapidly. Permeating our soils as never before are great quantities of solvent materials laced with benzene, chloroform, carbon tetrachloride, and most ubiquitous of all, trichloroethylene. Carbon tetrachloride, a simple molecule, is a potent carcinogen. Benzene, which is used in countless industrial processes, has been found to cause chromosomal damage at levels less than 10 parts per million. Inhaling such materials in the steam of a bathroom shower or ingesting them in a glass of tainted water can lead, at the least, to headaches or feelings of drowsiness; bathing in contaminated water can leave the skin cracked, red, and dry from the removal of its protective oils. Given exposure to sufficient quantities over an extended period of time, serious destruction of vital tissues is sure to occur. Sufficient carbon tetrachloride creates massive discharges of epinephrine from the sympathetic nerves that overwhelm the liver, leading to cirrhosis and damage to the central portions of its lobules. In the kidneys, where the poisons are ferreted out and sent to the urine, but where they may also collect and react, lesions may appear suddenly, or renal corpuscles may be permanently damaged. Internal irritations and scars may be followed by fatal growths of cancer: in experimental mice, rats, and hamsters, the administration of these agents has initiated malignant tumors.

The same can be said for many pesticides, wastes from which are constantly spreading through the water table. Taken into the digestive tract daily over a number of years, a quantity as paltry as 1 part per trillion, collecting in the milk of a nursing mother or in the fat of the intestines and other kinds of tissues, can do significant damage. And results can be quick. Dogs given water containing 160 parts per million of the chlorinated hydrocarbon toxaphene have developed perceptible changes in the kidney tubules and liver after forty-four days. Even when these substances do no direct harm, they still may damage the body's defense mechanisms enough to invite the incursion of infectious disease. In 1970 it was determined that ducklings fed pesticidal material and then infected with duck hepatitis virus died at a rate four times that of those unexposed to the synthetic agents. Some chlorinated pesticides, for example dieldrin and aldrin, can be expected to cause 1 extra cancer in a population of 10,000 if introduced into the water at less than a half a part per billion.

The effects of long-term exposure to small amounts of solvents, pest killers, and plasticizers have not yet been fully studied, and perhaps they never will be, for scientists cannot experiment on human beings as they do on rats. Yet our folly in waste disposal has begun inadvertently to create just that: an enormous laboratory that engulfs both cities and states, casting large numbers of people, without their knowledge, in the role of guinea pigs.

Some of the consequences of allowing humans to imbibe synthetic organic compounds were demonstrated in a study prepared in 1974 by the Environmental Defense Fund in Washington, D.C. Surveying the Mississippi River, the shores of which are crowded with industries and scoured by waste dumpers, the EDF noted that forty-eight chemicals had been identified in the river's waters, including chloroform, hexachlorobenzene, ethyl benzene, and dimethylsulfoxide. Several of them were making it into the treated tapwater, eluding the settlement in forebays and the sand of inadequate filters. While the contamination was increasing, so was the cancer rate among New Orleans residents. From 1949 to 1951, the EDF reported, New Orleans ranked third highest of 168 metropolitan areas in kidney cancer mortalities, sixth highest in bladder malignancies and urinary tract cancers as a whole, and ninth in all the country in cancers of the digestive tract. In a later survey, covering the years up to 1969, the New Orleans over-all cancer rate was 32 percent higher than the national mortality rate. For certain specific cancers, New Orleans's rates were three times higher than those of Atlanta and Birmingham, cities that had alternatives to Mississippi water supplies. Along the Hudson River, where PCBs have infiltrated the water supply, rates of rectal and colon cancer increased 65 percent in the city of Poughkeepsie from 1950 to 1970. Was it mere coincidence that in that twenty-year period trichloroethylene, carbon tetrachloride, and chloroform had also appeared in the river? And was it not logical that if these compounds were also in the ground, then cancer would begin to afflict those who thought themselves safe drinking from their own private wells?

For a quarter of a century, Salsbury Laboratories, a family-owned producer of veterinary pharmaceuticals, had dumped its waste concoctions on an eight-acre site that once had been a sand and gravel pit. The dump was along the Cedar River in Iowa, on a fractured dolomite and limestone formation that had developed subsurface water channels dipping to the southwest, and near to a small construction company, a bowling alley, and several small convenience shops that served the rural community of Charles City. Also in the vicinity, across a two-lane highway, were modest frame houses, mostly painted white or yellow, inhabited by neighborly people who felt the company had done well by the town despite a minor problem in the 1950s, when several private wells had been found contaminated.

Little more was said of the incident until 1969, when Samuel J. Tuthill, director of the Iowa Geological Survey, sent a letter to Iowa's water-pollution control commission after surveying the site. His words were formulated in a surprisingly strong fashion. Tuthill said he had "been made aware of a potential hazard to the citizens of the northeast section of Iowa." Because the underground channels acted like a hydraulic pumping system, Tuthill feared that solutes from the dump were beginning to move, at an unusually high rate of speed, into the Cedar Aquifer, which along with connecting flows supplied water to at least 302,000 people.

> Knowing what we now do about the geology of the dumpsite and the underlying rocks and having observation-well data to evaluate, I must recommend this problem as a very possible and serious hazard. The situation at the Salsbury dump is that arsenical, non-arsenical, and gypsum wastes are being delivered to an authorized disposal area between Highway 218 and the west bank of the Cedar River southeast of Charles City. There are six observation wells within the dump area and one on

> the east bank of the river. Periodic analysis of the water from these wells shows that arsenical compounds are highly mobile in groundwater.

A lot was at stake: between 1953 and 1977, nearly a million cubic feet of arsenic wastes had been buried there.

Tuthill's words went unheeded. Not only were no corrective measures taken to halt aquifer migration of leachate, but Salsbury continued to dump in the same place for eight years after Tuthill's warning. The company seemed to maintain that because there was no certain proof of injury or death from its wastes, and also because it couldn't afford the $20 million needed to permanently rectify the burial grounds, it should be allowed to proceed with business as usual. There were approximately 30 million cubic feet of sludge and contaminated soil, and no one in government or in the company knew what to do about it. There was lengthy discussion of moving the wastes to a sanitary landfill or to a new dumpsite with a more secure undermining, but those alternatives were never agreed upon as practical ones, and so the problem was simply ignored.

Upon further review of documents pertaining to the case, it became obvious that the EPA had been made well aware of the precarious circumstances surrounding the Salsbury dump at least as far back as 1972. But it too had chosen to take no action. It was on September 13 of that year that Don Marlow, of the hazardous waste management division, typed for the record a memorandum that should have raised some degree of alarm in Washington.

> These wastes are presently being stored in open dumps next to the company. In addition it appears that no one has performed a chemical analysis of arsenic wastes

> that are being stored in Charles City, Iowa. There are preliminary indications that the soluble arsenates are contaminating the groundwater supply in the area. Although additional evidence will have to be produced and reviewed to establish this fact. This situation could present an imminent threat to all living species in the area because of [*sic*] the underlying fractioned limestone bedrock is where 70 percent* of the residents of Iowa obtain their drinking and crop irrigation water.

Five years later, when the Iowa Department of Environmental Quality finally attempted to move into action by preparing an order to the firm to halt its dumping, it was nearly restrained judicially. A county district court judge, in a bizarre twist to a story otherwise without humor, issued a temporary injunction that forbade "issuing, delivering, giving, or transferring" any oral or written communication about the laboratory to anyone outside the department. The injunction thus restricted discussion of a major public health issue and temporarily prevented delivery of the cease-and-desist mandate. Soon after, an Iowa Supreme Court justice nullified the lower-court ruling; the order was belatedly issued, and Salsbury, announcing that it had to lay off sixty-five employees as a result, grudgingly stopped pouring its poisonous sludges into the dump and modified its operations as it looked around for a new landfill. What had provoked the state finally to take action was that traces of chemicals had appeared in drinking water taken from shallow wells in Waterloo—at least fifty miles from the disposal site.

When the wastes were analyzed, the implications of Salsbury's practices took on new proportions. An Ames

* Other reports put the figure at 10 percent.

test for mutagenicity, which gauges the potential of chemicals for causing chromosomal and general genetic changes in bacteria, had shown that the discharges indeed possessed this frightening capability. Benzene, orthonitroaniline, and twenty-two other compounds, many ranking on the EPA's list of "priority pollutants," were detected going into the Cedar River. In August 1978, after the Ames testing, EPA's enforcement division circulated a memorandum saying, "These results increase our concern about the LaBounty site, and reinforce the position of Region VII (EPA) on the need to isolate totally the dump from ground and surface water supplies. We believe a sense of urgency must be imparted to all concerned to ensure adequate protection of Cedar Valley water resources."

The memorandum was sent to the Iowa Department of Environmental Quality, which responded with remedial plans for sealing the dumpsite. But I was informed in June 1979 by Joseph Obr, director of DEQ planning, that the proposals from Salsbury merely included placing cover soil on top of the small dump to decrease leachate production, diverting surface runoff, and further monitoring the spread of contaminants. But, he added, "There really hasn't been any work of significance done at the site as of now." This ten years after the first warnings of danger. Even now Obr remarked that a consultant the state had once hired told the department there was only one solution: removing the wastes elsewhere. But there were no plans for that. Instead, thousands who drink from the aquifer will take the chance of consuming organic materials of high toxicity until a permanent solution is agreed upon between Salsbury and the various governmental agencies. For some, it may already be too late: according to the *Clean Water Report*, a preliminary study of people living in Black Hawk and Floyd counties, where Charles City and Waterloo are located, has shown what may be "elevated levels" of bladder cancer, but no proof of a connection.

There were alarming signs that the Charles City pollution, though perhaps unique in its breadth, was but one of many dozens of waste hazards that had been shelved by the regulatory agencies only to return in a later decade to haunt the environment and infiltrate the vital organs of plants, animals, and humans alike. It was a rare state that did not have a clear-cut example of well contamination for the EPA record books. I heard stories of livestock drinking from fouled ponds and turning sterile, and of families who had recurring seizures, rashes, and chest pains after ingesting mercury that had collected in their wells. The ailments continued to mount as landfills, excavated ten or twenty years before, began to dissolve their soil barriers, belatedly displaying their forces of debilitation. Eighty acres of groundwater in Saint Louis Park, Minnesota, had been tainted with phenols, threatening the water supply for twelve cities including Minneapolis. DDT, leaking from the United States Army's old Redstone Arsenal, had accumulated to the tune of thousands of tons in or near the Tennessee River. In Yerington, Nevada, there was a gasoline taste to the wellwater, with a community landfill located nearby. In Agana, Guam, old munitions dumps were suspected of discharging zinc into the groundwater. Then there was Hodgenville, Kentucky, where it was feared but not proved that paint thinners from a waste-collection facility had spread into streams and drinking wells; suddenly calves were reportedly being born with their heads and feet twisted, their tongues crooked, and their limbs uncontrolled, and without any hair.

Because of their isolated nature, farmlands had been

favored by many independent haulers as places to unload their unusual trash away from the view of authorities. They would trek down a back dirt road to pump, to dump, and then to flee responsibility. For an unknown number of years, hundreds if not thousands of 55-gallon and 30-gallon drums containing cyanides, heavy metals, and other materials from metal-finishing processes were discarded on farm property near Byron, Illinois, that was later purchased by the Commonwealth Edison Company for a nuclear reactor. The dumping had ceased around 1972, but its trail was observable in the contaminated runoff that appeared when rains were heavy. It was also to be noticeable in another way: on May 20, 1974, three cattle were discovered dead on the land from no apparent cause. The mystery was not so for long: pathological studies revealed death by cyanide poisoning. Looking for the cause, Commonwealth Edison found 1,511 containers of mire on what was known as "Dirk's Farm," and discovered that the seepage had damaged not just cattle but also waterfowl, vegetation, and the bottom-dwelling organisms of the closest stream. Several other dumpsites, unmarked and unauthorized, were found within a mile radius of the farm. In one test, water runoff contained levels of cyanide that were nearly two thousand times the federal standard for drinking water, leading to fears that the land in the general vicinity could never be used for crops again. While the surface flooding was poisoning the topsoil, far below the chemicals had contaminated the potable groundwater, so that wells serving sixty-eight rural homes had to be closed.

The same trauma befell the New Jersey property owned by the Samuel Reichs, a farming couple from Germany. About the same time the Byron case was unfolding, the Reichs had leased part of a chicken farm to one Nicholas Fernicola, on the assumption that he was in the metal-drum salvaging business and was to store empty barrels on their property until enough were gathered to deliver to purchasers. A few months later, Mrs. Reich told me, they inspected the leased land in an effort to find the origins of a nauseating odor lingering in the air and discovered that the barrels were not empty after all. They were drummed residues, strewn about and leaking. In December, the Reichs notified Mr. Fernicola of their find, and also the Union Carbide Corporation, which had contracted with Fernicola to dispose of its wastes—plasticizers, resins, and organic wash solvents. Receiving no satisfactory response from Fernicola, Union Carbide, or the state environmental agency, the Reichs took the case to court. Union Carbide was ordered to remove most of the drums.

Still, the problem persisted. During 1974, additional drums were found four miles away in a wooded area near the Winding River, and so was a trailer containing similar wastes. That same year, residents of the vicinity began to complain of the taste and odor of their tapwater, and when it was proved that petrochemicals had seeped into their groundwater, the township of Dover passed an emergency ordinance requiring 148 wells to be capped with cement.

Municipal dumps have also been a source of groundwater contamination. When poorly watched at night or infrequently inspected, such sites are often used for the dumping of industrial wastes. Drums of toxic residues, hidden in mounds of paper trash, are frequently transferred to a community landfill where nothing prevents the toxic substances from eventually finding their way into the groundwater. In other instances, city authorities, more concerned about their budgets than about the citizens' health, have allowed industrial wastes to be blended with city garbage, if the price was right. Despite warnings from the state that a proposed landfill would seep through crevices in the bedrock, the city of Aurora, Illinois, went ahead and used the area as a municipal dumping ground until 1965 and then sold it to a company that poured industrial effluents into a trench at the site. It was only a short time before several wells had to be closed because the community's tapwater, in the words of one visiting reporter, was "as black as the ink in this story" and smelled "as strong as a glue factory in midsummer." Hazardous wastes have been discarded at community trash heaps that were later covered over and forgotten. So it was that eleven people in Perham, Minnesota, developed arsenic poisoning in 1972 after drinking water from a well recently drilled by their employer, a local building contractor. The water contained up to 21,000 parts per billion of arsenic, quite in excess of the 50 parts per billion set as an acceptable level. The culprit was a mere fifty pounds of grasshopper bait, consisting of arsenic trioxide, bran, sawdust, and molasses, that had been buried by farmers during the 1930s in a corner of the village dump. While the amount seemed negligible, it was enough to cause severe neuropathy in one employee that cost him the use of his legs for six months.

The EPA once complained that 30,000 rural communities had unsafe or inadequate water and sewer systems, and that these accounted for many internal ailments, from tooth decay to cancer, across the United States. Ironically, one of the most serious affronts to a rural watershed was caused by the federal government itself, in a situation that, like Salsbury's arsenic, threatened the future of a vast amount of public water. At fault was the United States Army Chemical Corps (and later perhaps the Shell Chemical Company). At its Rocky Mountain Arsenal in Colorado, between the cities of Denver and Brighton, along the South Platte River valley, wastes from the manufacture of defoliants, pesticides, and chemical warfare agents had been stored in unlined canals and ponds in a geological region where alluvial deposits had acted like a water-table aquifer, allowing effluents to readily penetrate the ground. An area of at least thirty square miles, consisting of rich, powdery brown soil, had been poisoned as a result. The first indication of trouble had come in 1951, when beets, alfalfa, corn, and barley on property to the north of the arsenal, irrigated by shallow wells there, turned yellow and brown. Sheep and other livestock drinking from wells began to sicken and die. More than $160,000 had to be paid farmers for crop losses, and lawsuits were filed for alleged additional destruction resulting from the pollution of sixty-four domestic, stock, and irrigation wells. In a move to remedy the rampant pollution, the government spent more than $1 million on a deep-injection well to pump the wastes 12,045 feet below the surface and on construction of a 96-acre asphalt-lined storage lagoon. But the well had to be abandoned after it created earth tremors throughout the farmlands, and monitoring was to show migration even in the perimeter of the "secure" storage pit. So intimidating was the problem that, in April 1975, the Colorado Department of Health felt obliged to issue a cease-and-desist order against the army and Shell to stop pollution of the underground. No permanent solution yet has been implemented, and only partial decontamination is expected to cost a minimum of $78 million. In the meantime, wells sampled in one study by the United States Public Health Service have shown the presence of chloride, sodium, fluoride, arsenic, chlorate, and the insecticides aldrin and dieldrin. Concentrations of chlorate, used as the general indicator of contamination zones, were as high as 4,000 parts per million. It was no mystery how the chemicals had found their way into the soil and then through the miles of the underground watercourses. However, it was determined that one compound that had *not* been manufactured at the facility, 2,4-D, capable of causing the defoliation observed, was also present in the wells and in holding lagoons on the site. After a thorough inquiry, scientists decided that the 2,4-D had been created spontaneously in the open storage ponds, quite without the intervention of a chemist's knowing hands. All it had taken was the indiscriminate mixture of many compounds in one large earthen vat, together with exposure to air, rain, and sun. To make the matter more serious, it was found that the northernmost well indicating trace contamination was located only one mile south of the public water supply field for the city of Brighton.

Near the oldest landfills, the troubles were naturally most acute and severe. The Northeast, by virtue of its lengthier industrial history, had witnessed the highest number of well closings due to contamination, a harbinger for the South, the Midwest, and the Pacific coast. Indeed, it was fortunate that many Eastern states were served by fresh streams and lakes, for the rivers below the surface were deteriorating at an undiminished and uncontrollable pace.

On December 17, 1977, Robert McNally, health officer for the town of Gray, Maine, ordered sixteen wells capped after his office received notice that a young girl had been stricken with bladder and liver troubles. A few days before, McNally had informed the town council that well-water in one section had become so noxious that it should not even be used for flushing toilets because of fumes it released into the air. Tests conducted for the town by a

Cambridge, Massachusetts, engineering firm indicated that the wells were polluted by trichloroethylene and trichloroethane. The source was claimed to be the McKin Company, a small business that had begun by cleaning oil tanks but had expanded its commercial horizons to the collection and reprocessing of waste oils and solvents. Corroded storage tanks and ponded wastes were found in a sand and gravel pit on the site, and seepage was detected 600 feet below the surface.

Gray officials voted to extend municipal water lines to the stricken homes, but there was a period when it was feared that such a move would only feed in more trichloroethylene. The degreasing compound had also been tracked, in trace quantities, in the general community water supply, presenting residents with orange-colored tapwater. It was enough of a health concern for 750 families, in January 1978, to be ordered to stop using their water for cooking or drinking for a full week until the levels had receded. Reports circulated throughout Gray of skin rashes, loss of equilibrium, and more liver disorders. Legal actions were filed against McKin, which initially refused to spend the money needed for a housecleaning, and the state health department warned municipalities drawing water from the Royal River that the thick leachate might also have reached their intakes via the underground routes.

Trichloroethylene was also to blame for well closings in Canton, Connecticut, a suburban community that prides itself on its pristine country drives and its distance from car emissions and industry. During the summer of 1969, according to the testimony of a former employee, about 40,000 gallons of waste chemicals were dumped on the ground near Route 44 by a chemical company in the business of distilling residues. Barry W. Cosker, who had been employed by the John Swift Chemical Company, recalled to the *Hartford Courant*: "There were a whole bunch of barrels out back that were old and had been there from before my boss had the company. I didn't know what was in the barrels and neither did my boss." They knew only that they had to get rid of them.

Some major problem areas have drawn contributions from multiple polluters over many years. More than one hundred years ago in North Woburn, Massachusetts, near Boston, Merrimack Chemical began producing chemicals for the textile industry, and later expanded its facilities until it also became one of the largest producers of arsenic-based insecticides. In subsequent decades, other firms operating on or near the site—including the Monsanto Chemical Company and the Stauffer Chemical Company—produced animal glue and grease and other environmentally damaging items. It was Merrimack, however, that created the most significant stockpile of residues. In the summer of 1979, officials learned that there was an open pit, a dry lagoon, covering about an acre of land in which arsenic was piled in caked white powder several feet thick. So concentrated were the arsenic, lead, and other chemicals that a mere 45 pounds of the soil would be enough to administer a lethal dosage to 100 adults. That same year two nearby municipal drinking wells had to be shut down because of trichloroethylene, while other underground watersheds remained in severe jeopardy because of the forgotten dumpsite. Authorities have already requested blood tests for workers at an industrial park on the site, and for local residents. To compound the problem, the arsenic has been scattered widely by winds and rain and has contaminated a river watershed known as the Aberjona, which courses through Winchester on its way to the Atlantic Ocean. Nearby Mystic Lakes may also be endangered.

Suburbs of New York City, especially on Long Island,

were confronting a growing chemical menace beneath the ground. Most of the underground water system on the island extended as deep as 2,000 feet and was relatively free of toxins, but the top 100 feet or so, to which many wells were connected, were extensively polluted. Four-fifths of the shallow wells tested in one government study contained traces of trichloroethylene, and a sampling of 450 public water wells in Nassau County showed that 25 percent had traces of suspected carcinogens. The levels of organic contaminants were generally higher than elsewhere in the nation (with a few exceptions like New Jersey). From Farmingdale to the South Shore, a large subsurface mass of chromium, used in World War II aircraft construction, was slowly working its way to the Atlantic Ocean, closing wells in its path. In Glen Cove, half of the community's wells were shut when levels of trichloroethylene three times the acceptable limit were reached. Twenty wells were closed in Bethpage and others in Roosevelt, Syosset, and Garden City Park.

In another New York suburb, West Nyack, I talked to Mae Costa, whose well and whose neighbors' wells had been contaminated with trichloroethylene. She related: "My nextdoor neighbor called us over to her house one night and asked us to smell her hot water. It smelled up the whole house. It was like rotten eggs—horrible! The Rockland County Board of Health came in and tested—it tested for bacteria! For two years, I told my husband our water smelled funny, but he said it was only in my head. And here they had tested for bacteria and that was fine. When all of this broke, it was a Sunday and the board of health called telling me not to drink the water and to tell my neighbors not to drink the water either. But what was happening is that they wouldn't tell me where it was coming from. I called the Nyack Hospital Poison Control

Center and asked the side effects. And they said headaches, nausea, vomiting, and that it could lead to heart disease and kidney disease and cancer. With the side effects, I went to my neighbors. We all sort of panicked. They wanted to bring us in water at our expense, but I wasn't about to pay for something that wasn't my fault. So they got a grant from the federal government and also money for medical tests—physicals. One neighbor went back for four or five blood tests. One I think had spots on the liver, another white corpuscles, a high count. They said they saw no problems related to the water, though. But in my husband they found a heart problem and he's only forty-two. Me, myself, in 1976 I was hospitalized for very severe pains in my head. I couldn't remember my own name. I spent a month in the hospital and they couldn't diagnose it. My own doctor didn't know what he was looking for. I had a Pap test that was a 'four.' Sooner or later, my doctor said it could turn into cancer. One day we sat down with all this and thought about it and how everything had happened since we moved into this house. We aren't drinking the water since then, but what if we get cancer fifteen years from now? Who pays for that? What about my daughter? They told me they want to follow her the rest of her life to see what effect it has on her and her children. You think of that and after a while you learn to force yourself to smile."

There are several theories on how carcinogens, blended with our drinking water, cause cancerous growths. One older hypothesis emphasizes cellular oxidation, the vital process whereby minuscule components of a cell, including the mitochondria which contain packets of enzymes, fragment and chemically react with the glucose fed through the membrane, arranging and rearranging phosphate groupings in a never-ending cycle that produces the energy. All this is done in the tiniest of workshops: some cells placed 4,000 side by side would constitute a row only an inch long, and the mitochondria, of course, are even smaller. Should a synthetic molecule find its way undestroyed into the cell, combining with the fat or touching upon the mitochondria, it can inhibit the initial glucose breakdown or deactivate one or more of those enzymes so necessary as chemical controllers in the process of cell respiration. Or it may uncouple phosphate groupings in a fashion that inhibits the normal manufacture of energy. In short, such an interruption would portend damage to a cell or group of cells, in turn perhaps causing enough damage to impede the normal functioning of a tissue, and if the damage is still greater, the healthy workings of an organ. If it does not completely destroy the cells, this chemical wrench jammed into the machines of oxidation may force the cell to seek other means of nourishment. One way is the primitive one of fermentation. This abnormal cellular activity continues on through cell divisions until a group of cells are transformed in both function and appearance to the point where a cancer is said to have developed.

A more popular hypothesis is that the synthetic compounds alter or even destroy the proper functioning of a cell's DNA and chromosomes, which carry the genetic memory and play an active role in cell division; when they are damaged, so too is the manner of cell division. Less directly, the chemical agent may deactivate the body's built-in control against cancerous proliferation, allowing the erratic cell, perhaps triggered by another carcinogen, to continue on its wayward path, dividing time and again until it forms a demented mass that starves those tissues next to it.

Until the exact mechanisms of carcinogenesis are known, we can consider no quantity of a potentially cancer-causing agent in our water to be a safe one. We cannot predict from the size and substance of a cell which compounds might spark a malignant outburst. Just as a cancer may be initiated through a single cell transformation, so a mere 1 part per billion of a carcinogen, or even 1 part per trillion, may be sufficient for a full-blown malignancy. In fact, small doses, because they do not kill cells outright but instead mutate them, may be more dangerous than a large "hit."

For these reasons there is no "safe" level for ingestion of carcinogenic materials. Yet our government agencies continue to establish arbitrary quantities below which the intake of solvents and pesticides is considered safe. The way they do this, I was told quite bluntly by a member of EPA's human effects division, "is that somebody just pulls a figure out of his ass; there's no real scientific factor." In actuality, there is a basis behind the tabulations, but it is a dubious one: the levels are established through the extrapolation of observed chemical effects in laboratory animals. But such a determination is questionable for a number of reasons. Effects may vary from animal to animal, even members of the same brood, and certainly there are differences in the actions of residues upon rats and humans. A larger animal has a greater number of cells that may be vulnerable to the havoc wreaked by chemical agents; therefore the smaller animals used in laboratory experimentation may be less susceptible to damage than man. Thus, an ad hoc committee to the surgeon general in 1970 proclaimed that "no level of exposure to a chemical carcinogen should be considered toxicologically insignificant for man."

Still the regulators continue to establish "safe" ingestion levels; still they allow for the burial of carcinogens near the groundwater. In case heaped upon case, the authorities have failed to monitor drinking wells located near chemical dumps until residents repeatedly complained of water that smelled like turpentine or hair spray, or like the fumes from a dry cleaner's shop. By that time, the people may already have consumed enough groundwater to provoke a cancer that, years down the line, will end their lives. The United States Water Resources Council expressed concern about this dire lack of information in a 1977 Nationwide Analysis Summary Report: "In all the regions studied, there is a prime need —to mount a concerned effort toward locating, monitoring, and evaluating other existing cases of groundwater contamination, and to determine the effect of this contamination on our health before we reach the point of no return that some other parts of the world now appear to be reaching." Unfortunately, there was hardly a local or state official paying any heed to statements like that of the council.

6

TENNESSEE: "NOBODY CARES NO MORE"

The shortcomings of local and state governments in dealing with the irresponsible waste disposal procedures of industry were illuminated in a trauma that befell a tiny community seventy miles east of Memphis, Tennessee, among the cotton fields and clannish cabins of Toone-Teague Road near the town of Medon. Along a sizable stretch of the unevenly paved course, the Velsicol Chemical Corporation, based in Chicago but with an important plant in Memphis, had hauled its waste to shallow pits and trenches on a 242-acre tract, beginning in 1964. The area, to be sure, was sparsely populated, a dozen or so homes in the immediate vicinity of the dump and several more up the road. They were inhabited by a country folk who lived from day to day, working on a small construction job here, or as a sawyer in a mill there, shooting quail and hillbilly rabbits occasionally for their dinner. Bobcats and deer roamed through the clover and tall grasses.

In an area of small gullies and ravines, above three strata of groundwater and a sand-clay mantle, Velsicol had implanted 300,000 55-gallon drums and fiber cartons filled with residues. They were buried in trenches on three plots, each trench approximately 12 feet deep and 15 feet wide, and topped over, it was said, with 3 feet of soil. The territory was an upland remnant of a fluvial terrace, the clay bright red and the sand everywhere to be seen. The uppermost 250 feet embraced a "perched" water zone and water-table and artesian aquifers. Thus the materials were set upon a copious supply of groundwater and in earth not equipped to contain large volumes of leachate.

Velsicol's products—and wastes—were similar in nature to Hooker's chlorinated hydrocarbons: cyclodiene pesticides such as heptachlor, endrin, and dieldrin. These materials began to be produced in large volumes by American firms in the 1940s, with the diminished use of naturally occurring organic and inorganic agents such as the arsenates, mercury, and rotenone. Cyclodiene sales too were eventually to wane, for three of the products were banned or otherwise restricted because they had been found to cause cancer in laboratory animals and, like dieldrin and its sister compound aldrin, were discovered to have such a cumulative effect on the food chain that once their presence in the flesh of fish in a Michigan lake was detected at nearly a billion times the level present in the lake water itself. Before the restrictions, cyclodiene pesticides had been produced in vast quantities, perhaps as much as 65 million pounds in some years, and now their residues, tucked under the ground, continued to menace the population.

Indeed, each of these agents was renowned for its persistence in the soil and its sharp toxicity to animals. Sprayed upon Midwestern fields, they left robins' eggs lifeless, squirrels and muskrats dead in their tracks, and cats all but eradicated. In 1959, heptachlor was blamed for destroying virtually all the birds and much of the other wildlife over a 300,000-acre area near Joliet, Illinois; in 1957 it was used to kill fire ants in Louisiana, and did so with such efficiency that the insect's natural prey, the sugarcane borer, proliferated out of control, causing serious crop losses. These traumas were long ago reported upon by

Rachel Carson. Once in the soil, where it can be detected as long as ten years after a surface application, heptachlor is known to transform itself into a more toxic agent known as heptachlor epoxide. This agent too is fat-soluble and therefore likely to accumulate in the food chain. A Food and Drug Administration test in 1952 found that female rats fed 30 parts per million of heptachlor had stored 165 parts of its epoxide in only two weeks' time, a level higher than that found to induce cancer in laboratory rats. Furthermore, heptachlor epoxide has been detected in human mother's milk and in the blood of the umbilical cord, from where it could attack the tender tissues of the growing fetus. Aldrin and dieldrin, once so widely employed to control corn-soil pests and termites that they ranked sixth in sales among insecticides, were manufactured by the Shell Chemical Company and handled by a multitude of firms until they were banned in 1976 as an imminent carcinogenic hazard. Absorbed through the skin, dieldrin is perhaps forty times as toxic as DDT; aldrin, in a quantity the size of an aspirin tablet, is capable of killing hundreds of quail. Endrin in certain circumstances is more toxic than all the others, harming shrimp at levels below 1 part per billion. In humans these toxicants can cause afflictions as seemingly innocuous as insomnia and nightmares or as devastating as convulsions and brain damage. They enter the body through food and water, through the skin, and by riding particles of wind-blown dust that make their insidious way down the trachea and bronchi to lodge in the air sacs of the lungs.

Three years after the first drums were hauled to Toone, the United States Geological Survey's water quality division reviewed the landfill in a forty-page report entitled "Potential Contamination of the Hydrologic Environment from the Pesticide Waste Dump in Hardeman

County." Stamped beneath the title were four words: "For Administration Use Only." The report was written, according to its introduction, when "public health officials became gravely concerned about the potential health hazard owing to the toxic nature of the wastes" and "the possibility that these toxic wastes might contaminate, and then render useless, local and contiguous water supplies." The survey generally gave the impression that because they were situated upgrade from the dumpsite and away from the prevalent northeastern groundwater flow, the wells on Toone-Teague were not in immediate danger. At the same time, it added that "a substantial portion of the local water-table aquifer, the principal source of local domestic water supplies, is exposed to the hazard of contamination from the pesticide disposal pits." It was already known that the "perched" water body had entrained some of the toxicants and that the surface was polluted. Analytical results showed the presence of heptachlor, dieldrin, and heptachlor epoxide in the sediment washoff or in the bedload of Pugh Creek, more than a mile away. Still, the report concluded, there was no evidence it would soon reach the wells.

Those officials who reviewed the study apparently took it as a reassurance despite the indications that widespread contamination was beginning to occur. Consequently, the dumping was allowed to continue for five more years, and no one adequately monitored the drinking wells.

In February 1972, a full fifty-three months after the study was issued, Velsicol was finally ordered to stop dumping. This was at the behest of Dr. Eugene W. Fowinkle, state health commissioner, and S. Leary Jones, director of water quality control, who, concerned about the residents located immediately west of the landfill, argued that "underground conditions and groundwater movement directions are not known in these areas" and "groundwater could move westward and contact water supplies." They continued: "Below the local aquifer is an artesian which slopes and flows westward. This aquifer is used as a water supply by many West Tennessee cities, including Memphis. . . . The second danger is presented by the proximity of the southern boundary of the dump to headwaters of a stream which flows southwest through the town of Toone, Tennessee."

State officials were deluded if they believed their order would have any effect on the leachate below, which was moving at a rate of perhaps eighty feet a year and heading, in one of its paths, toward the private wells on Toone-Teague Road. Around Christmastime in 1977 Steve Sterling, a resident of the road whose tapwater had been tasting and smelling strange, took a sample of his well-water to the county health department, where he encountered George Wallace, the environmentalist. Wallace recounted to me that he had known of the Velsicol dump for years. One wonders, then, why Wallace, a man with a degree in chemistry, at first checked the sample only for bacteria. He explained: "I never heard about chemicals like that. I did everything I know to do. I sent it to the state and they couldn't find anything in the parts per million." But upon going to the area and smelling the water, Wallace said he noticed "something was in there." Mr. Sterling, as well as some others on the road who had complained persistently, remembered it this way: "He [Wallace] didn't even want to test it. It was like there was nothing wrong with it. They left that impression. I carried a sample over there twice. It seemed he was always hiding when I did. He said he would be there, and then I'd have to wait two hours. He said he didn't have the time to come and get it himself. He said he couldn't smell anything, but you could. He just didn't want to admit it; he wouldn't taste it, wouldn't drink any of it, you see."

While the county and state were not quite rushing to an investigation, residents were beginning to fall ill. Steve Sterling became so fatigued that he could walk across his yard only with the greatest difficulty. One of his daughters had kidney problems, and his wife was hospitalized for respiratory ailments and chest pains. Others along the road were also feeling peculiar pains. John Boyd, a tall slim man with a craggy face and jutting chin, was riding home from Memphis in March 1978 when first his left hand and then his left leg became paralyzed. At the hospital they said it was a stroke, but no further explanation was offered. When John and his wife, Ethel, gave more thought to their health, they realized that at least since 1977 there had been innumerable discomforts in the household—dizziness, loss of hair, kidney pains, and coughing. "You couldn't sleep," Ethel said. "I mean, you'd get up and it would seem like you're going to fly off. Everything would go around and around. I can't explain it. Then I would vomit." Their daughter, Christine, who had married Woodrow, one of the Sterling men, became partly paralyzed on her right side. After a bath, residents needed to soothe their bodies with baby lotion, for the water made their skin dry and chapped and caused pimples. Smells like those of a factory rose from the hot-water boilers. Chlorine bleach poured into a washtub would turn the water brown. Yet they were reassured that all was well. Mr. Boyd claimed: "We started a'calling the health department. They said it wasn't anything, wouldn't hurt us. They would hand us stuff and they would say, 'There's nothing in it.' I could go down there and look him [a health official] right in the eyes and he'd say, 'It's the Sterling well that's polluted.' To the Sterlings he'd say,

'It's the Boyd well that's bad!' " Christine said she was angry because when she called the regional EPA office in Atlanta she was told "we were a low priority because we weren't a public well."

Into the spring of 1978, the state, having discovered chemical traces in the wells, first warned the residents not to drink the water, I was told, then somewhat reversed itself, then again said the water should not be consumed. Even before that, on the advice of Citizens Against Toxic Herbicides, an environmental group based in Clarkston, Washington, some of the residents had already halted use of their wells, hauling in water from outside the community, taking their wash to a laundromat, and bathing at Chickasaw State Park—far enough away, said Mr. Boyd, "that you had to stink a little to make that bath feel good when you got there." There was no public water they could use for their own homes until the summer of 1979. In the meantime, a better program of monitoring had detected chloroform, benzene, heptachlor, and chlordane in Woodrow and Christine Sterling's well. The carbon tetrachloride level, according to a congressional subcommittee staff that investigated the matter, was 4,800 parts per billion, "2,400 times the maximum daily exposure suggested by the National Institute for Occupational Safety and Health as safe for workers." It was only after Christine testified before the subcommittee, on October 30, 1978, that the EPA, in a certified letter, warned the family to "refrain from using your wellwater for any human contact."

"We would have been waiting on EPA forever," said Christine. "I'm very hurt and confused of their not taking any more interest in this than they have. They have really hurt us."

So embittered were the Toone folks toward all levels of

government that they began to favor the guilty corporation, Velsicol, over the regulators and legislators. Velsicol had done more to bring them relief than the governmental agencies. It had installed a temporary storage tank for the thirteen homes most seriously affected, had offered to buy the houses (a deal at least two families accepted), had replaced contaminated porcelain, hot-water tanks, dishes, plumbing, and washing appliances, and had paid settlements for property damages and inconvenience. "We think what we've done in Hardeman County should be a case study," Richard Blewitt, Velsicol's vice-president of public affairs, told me. "We set in motion a plan to rectify the situation and initiated press data." He forgot to mention that Velsicol had continued the dumping years after the first indications of environmental contamination, and then stopped only because it was ordered to. Nevertheless, as Christine Sterling said, "Velsicol was the one that brought in fresh water. No one else was there to help."

That anyone concerned with the environment could feel an affinity for Velsicol was ironic indeed. Aside from Hooker, no other company had been involved in as many known ecological abuses: its record ranged from a fish kill in the Mississippi to illnesses among plant workers in Texas. In Michigan, flame retardants from a firm it purchased had found their way into livestock feed, leading to the destruction of animals across the state. In Memphis, it was suspected that toxic emanations from its plant had caused employees of the municipal wastewater treatment plant to fall ill. As an official of the Environmental Defense Fund put it in an interview with the *Wall Street Journal,* Velsicol, a part of Northwest Industries, was a "corporate renegade" that seemed to be "totally lacking in public responsibility."

Despite public rebukes, Velsicol went to great lengths to continue making hazardous materials. The company once refused to comply with a federal request to call back two chemicals, and in 1976 it was publicly chastised by a United States Senate committee staff for permitting the further production of a pesticide, leptophos, after it appeared obvious that workers were becoming ill in its vicinity. At the same time, there were allegations that Velsicol had withheld information on another product that a company medical consultant had asked the firm to stop manufacturing because of the dangers it posed.

Controversy flared highest around Velsicol's manufacture of chlordane and heptachlor. Since 1965 heptachlor, along with its epoxide, had been suspected of causing cancer in mice. Velsicol claimed otherwise and contracted with the Kettering Laboratories of the University of Cincinnati in Ohio and the International Research Development Corporation of Mattawan, Michigan, to study the effects of chlordane, heptachlor, and the epoxide. While it seemed to these hired researchers that there was a dose-related incidence of "liver nodules" in the animals, in essence they deemed the compounds noncarcinogenic. A government review of the liver sections, however, declared them infested by cancer. Later it was learned that two consultants to Velsicol had reviewed some of the liver specimens and, in a letter to the company in 1972, had reported them to be cancerous. Five years later, on December 12, 1977, the United States Attorney's Office for the northern district of Illinois announced that a special grand jury had returned an eleven-count indictment against the company and six of its past or present officers, employees, and attorneys, charging that from August 1972 to July 1975 the defendants had conspired to defraud the government, concealing material facts indicating that the substances induced animal tumors. The charges were later dismissed.

Still, no one knew the extent of physical injury, and no

governmental agency seemed anxious to find out. A task force composed of state and federal officials decided against an extensive evaluation. "Since there has been limited exposure, it was decided unwise at this time to pursue any more specific individual testing for toxicity," said Dr. David T. Allen, deputy commissioner for the state's medical services. He added that there had been no unusual outbreaks of disease, no consistent pattern of illness—a viewpoint that was plainly misinformed and superficial. There were, after all, consistent reports of headaches, limb numbness, kidney troubles, respiratory problems, and nausea.

In yet another ironic twist, when a medical review was finally effected, it was done by the Kettering Laboratory of Ohio. A press release issued by the laboratory on June 21, 1979, concluded: "This study was performed in November of 1978 and about 25 percent of the blood tests that indicate liver function were above the normal limit in the 39 people who were tested." That was "significantly higher" than a control group in Memphis, and although conditions improved later on, it was further cause for wonder at those public officials who had seen no reason to study the problem. During subsequent liver examinations, seven individuals were found to have borderline enlargement of the liver, while only one person in a nonexposed comparison group displayed the same symptom. I was told by Christine Sterling that one of the Kettering researchers had called to inform them that the EPA was resisting the proposed study and that she was therefore asking their permission for a medical review. Despite repeated inquiries from the residents who ate the land's wild game, government technicians apparently had also neglected to inform them of possible contamination of the animals. When the family of Thomas Young sent one of its hogs to the University of Vanderbilt for an evaluation, tests showed its tissues to be contaminated by chlordane, and it was recommended that the animals not be butchered for food. Trace levels of endrin and other substances began to show up in blood tests of the human population. The local physicians were of little help, informing the residents they did not want to become involved because of the time it would take to appear in court. When Steve Sterling's physician was asked if the family's troubles could be from the water, Mr. Sterling said the doctor replied that "he wouldn't say so even if he knew."

There are plans now for remedial construction on the Velsicol dumpsite, and while some families continue to drink the water drawn from wells along the road, the public line was to be connected as of July 1979. But the damage, psychological as well as physiological, has been done. "Before this happened," said Ethel Boyd, "everybody was happier." The people had felt better about life, her husband added, but the dumping incident and the sluggish government response "knocked the props out from under us. We felt let down." The disenchantment was also apparent in the Boyds' twenty-six-year-old son Raymond, who lived next door. On April 27, 1978, Raymond's wife, Ona, gave birth to their first child, Benjamin. At his premature birth, the boy was little more than a skeletal figure who demanded immediate and serious medical attention; part of his digestive tract had formed on the outside of his body and had to be surgically tucked back in. Benjamin survived, but Raymond's attitude toward his society became morbid. "They said it wasn't heredity," he commented in his Southern drawl, "but can't get no one to say it was chemicals. Can't do it. They're all afraid, and seems like, like nobody cares no more anyhow."

7

NEW JERSEY: BLIGHT IN THE GARDEN STATE

Because of its truck farms and its patches of buttercups and violets, lush along the gently sloping lowlands, New Jersey is known as "The Garden State." But the fifth smallest and the most densely populated state less proudly lays claim also to the most serious ground pollution problems in the nation; according to United States Senator Bill Bradley, who has been a crusader on the issue of hazardous waste contamination: "Nowhere is the problem of toxic waste cleanup more critical than in the state of New Jersey." Its location near major sea routes and large urban centers has been a factor in its ascendancy in heavy industrialization, and for many years the manufacture of chemicals, involving 125,000 jobs in more than a thousand factories and laboratories, was heavier in New Jersey than anywhere else in the United States. One consequence is that the state is surpassed by no other in its oversized landfills and contaminated drinking wells. Another is that its piedmont region, in the north, has become the widest lane of a grueling industrial stretch known as "Cancer Alley."

The New Jersey Department of Environmental Protection issued on December 26, 1978, a press release headlined DEP'S GROUNDWATER STUDY SHOWS GOOD WATER QUALITY. Summarized in the glowing announcement were the results of an ongoing, multiyear water-testing program the department's Program on Environmental Cancer and

Toxic Substances had initiated after it was learned that New Jersey had the highest cancer incidence in the country between 1950 and 1969. It was somewhat perplexing, therefore, to read that the main finding was that "trace levels of some of the chemicals under investigation" were contained in "virtually all the wells tested." The study in question included 163 wells in nine counties.* Moreover, after the preliminary testing three public supply wells—not previously cited as trouble spots—were taken out of service because of the excessive presence of trichloroethylene and trichloroethane, two compounds that belong to a large family of solvents used as degreasers, as refrigerants and fumigants, in organic synthesis, and in the dry-cleaning business. The unabridged technical report, the basis of the press release, contained a warning: "Clearly, from our analysis of groundwater data so far, the findings of low levels of chlorinated organics throughout New Jersey demonstrates the imperative need for close follow-up and concerted effort to remove sources of contamination to the state's groundwater. Particularly disturbing, among this set of wells, was the finding of contamination of two public drinking-water supply wells in Camden County with trichloroethylene and one in Middlesex County with trichloroethane."

There were other remarks in the more detailed report that seemed at odds with the cheery perspective of the official press release. In two wells, the metals copper and chromium were found in amounts above federal water standards while the state was testing for nine elements in this category. The effects on human health of the metals analyzed are widely varied. Copper and zinc, in small amounts, are necessary for human metabolism, aiding the

* Atlantic, Burlington, Camden, Cape May, Gloucester, Mercer, Middlesex, Ocean, and Sussex.

function of enzymes. On the other hand, arsenic, beryllium, cadmium, chromium, and nickel have been linked to increased incidence of cancer in cases of occupational exposure and have clearly demonstrated carcinogenic potential in laboratory animals. Besides solvents and metals, quantities of pesticides such as lindane and heptachlor epoxide were also tracked. "Even though concentrations were very low, it is a continuing cause for concern that in approximately sixty instances, chemicals in this category were detected in well samples," said the report. "Polychlorinated biphenyls (PCBs) were found in thirty-two wells, the most common occurrence for this group of chemicals. . . . Although, for this set of wells, the concentrations of pesticides and related compounds were low, it is unfortunate that these compounds should occur at all in our groundwater supplies. . . . It should not be a cause for complacency that in the remaining instances for potable wells, concentrations did not exceed recommended levels." In a previous test of 250 wells, twenty violations of federal standards were found at nonpotable water supplies, the source of pollution unknown.

In the offices of the state's water division in Trenton, a pin map of New Jersey was demarcated with black circles for landfill seepages, red for drainage from industrial pits and lagoons, and several other colors for subterranean migration. Well in excess of a hundred circles crowded the map. If it is kept up to date, it will soon be too cluttered to read. As many as 3.5 million people may be affected. Many of the groundwater problems have occurred in Passaic and Bergen counties. A reporter for the *Bergen County Record*, a local newspaper which has kept close watch on the area, told me that in one five-year period between 10 and 15 of the state's public water systems had been closed because of poisoning and 500 private residences had been affected. Wells had been closed in Allendale, South Brunswick, and Ocean County, and also in Camden, where levels of trichloroethylene were found, at nineteen times the level New Jersey deemed safe, in a wellfield that had to be kept flowing because cleaner water was not readily available. In the well-kept middle-class community of Fair Lawn, near the state's northeastern border, two wells were closed in the fall of 1978 when toxins at levels more than one hundred times government standards were detected, and the following January, with no public explanation, the town suddenly shut down two additional wells and days later put all of the community's fifteen wells out of operation while testing progressed. Such action had not been taken months before, the city manager explained, because officials were waiting to see if the levels of chemicals would recede. Chloroform and other potential carcinogens were found in the thousands-of-parts-per-billion range, but secondary sampling showed, in one well, a level of 136 parts per billion of trichloroethane, far lower than the 10,940 parts per billion found previously by an independent laboratory. The startling discrepancy was explained away by authorities as a probable mistake in the first laboratory's analysis, but it seems more prudent to assume that at times the contamination had been much more serious—that the levels dramatically fluctuated from one day to the next.

Local governments not only tended to minimize the risks of groundwater problems but had at times even been blamed for causing them. In Jackson Township in Ocean County, a landfill owned by the town was cited for badly tainting the aquifer with acetone, benzene, trichloroethylene, and toluene. The landfill was not supposed to accept chemical wastes, and yet solvents had apparently been injected into the waste inflow either without the knowledge of the town guardians or without protest on their part. The Jackson landfill had been constructed on an area that had been mined, and so there was little topsoil or clay, making it simple work for the chemicals to reach the water table only forty feet below. Residents within a three-mile radius of the landfill were advised against using any of the 140 wells in the zone. There were numerous complaints of skin rashes, stomach ailments, and kidney irritations among those who had drunk from the wells, but when I asked a health official, Robert Gogats, about these alleged illnesses, he responded, "I think we caught it in time. Rashes are a nuisance and there was some diarrhea, but they did not drink these chemicals to any great extent, and it was explained to me that certain people have a tolerance to certain chemicals. These people have been using the water and they built up a tolerance." Those who lived there, however, said the illnesses were much worse than Gogats reported.

Neither the state nor the water companies were sure how to handle the massive problem, which had also caused more than 100 wells in Perth Amboy to be closed. Those company representatives I spoke with took the stand that the contamination problems were minor and the federal standards too strict. Acknowledging nevertheless that some action had to be taken, in March 1978 the Hackensack Water Company began a chlorination process to reduce levels of cancer-causing chemicals, using a chlorine-ammonia compound to lower the levels of trihalomethanes in the water. There were problems with that procedure, however. The process had to be halted when pet fish exposed to the water began to die mysteriously. For its part, the state told the press that it would probably be the 1980s before its environmental watchdogs could get "a handle" on the extent of contamination and be able to determine safe levels. This timetable was not attractive to State Senator John Skevin, who, according to the *Bergen Record*, said that unless there was prompter action the situation "would be likened to using New Jersey's citizens as guinea pigs, with their cancer rates providing the final statistical data on whether or not these substances really cause cancer."

Citizens in New Jersey were not pleased to learn in 1976 that during the 1950s and 1960s their state had had the highest white male cancer mortality rate in the country and had also ranked among the highest for females and nonwhites. While the state had 3.5 percent of the country's population, it had produced a full 4 percent of new cancer cases, 14 percent above the average. In raw numbers: Each year, 14,000 people in the state died of "environmental" cancer, 24,000 were in the early stages of developing it, and 1 of every 4 New Jersey residents, throughout a lifetime, would be touched by the insidious disease.

When the figures were further evaluated in 1979 by Dr. Michael R. Greenberg of Rutgers University, they clearly showed that the urban-industrial corridor between New York and Philadelphia exceeded expected fatality rates in all forms of cancer by about 20 percent, with particularly high rates for lung, stomach, female breast, rectal, and intestinal malignancies. Some of the neoplasms could be blamed on ancestry; for example, Poles, Italians, and Russians seemed more prone to pancreatic cancer. Some were said to result from exposure to automobile fumes and the rays of the sun. Some were attributed to personal habits such as tobacco smoking and diet. But when the State Senate Commission on the Incidence of Cancer issued its first interim report, it made clear what it thought to be a major cause: "There is a correlation between the high incidence of environmental cancer in New Jersey and its status as a manufacturing center, particularly of chemicals and related materials." Further, the commission took the position that "the principle of 'innocent until proven guilty' applies to persons and not to chemicals. Chemical substances should be judged guilty until proven innocent, with the burden of proof on the chemical and the benefit of the doubt extended to the people."

As with most studies of the sort, few definitive answers were offered and most of the cancer correlations, however obvious to the glance, were not quite statistically significant. But the maps that plotted the disease consistently showed its course as intensifying along the industrial spine, especially in Hudson, Essex, Middlesex, and Passaic counties and areas immediately adjacent. Another report, circulated within the DEP, cited toxic wastes as one of six major categories that had to be addressed if the cancer rate was to diminish appreciably. "The treatment and disposal of toxic, hazardous, and carcinogenic wastes presents a serious problem for New Jersey," said the summary. "With its large concentration of industry, New Jersey produces a large quantity of such wastes annually, and has, in addition, accepted for disposal large quantities of such wastes from other states. Historically, these wastes have not been treated before disposal; they have been disposed of either through ocean dumping or in sanitary landfills. In the past, little was done to prevent these materials from being washed out of the landfills into ground and surface waters of the state."

The state's response was to organize a "cancer registry" program, propose bans on certain air emissions with heavy fines for offenders, and establish a cancer control council. Said the senate report: "Environmental cancer represents a clear and present danger in New Jersey; a threat not only to the lives and happiness of our current adult population, but to our children, their children, and those yet unborn."

Eighteen months before the senate commission issued its dire warning, Vivian Cleffi, a young housewife in Rutherford, New Jersey, near the matrix of highways and hamburger drive-ins that dominate these Jersey locales, was told by a New York City doctor who had examined the blood of her eldest son, eight-year-old Jimmy, that the boy had leukemia of the very worst kind. On September 6, 1976, at 2 A.M., his large brown eyes long since blackened underneath, his stomach swollen, a hole in his throat where a tracheotomy had been incised, Vivian's son quietly curled into the fetal position and died.

Although those who knew the family were naturally saddened, initially the death had no special effect upon the 21,000 in the community, many of whom lived in the large wood and brick homes that abounded in the town, under the shade of huge elms and lindens. Rutherford's calm state, nurtured by its isolation from the main thoroughfares, was soon to change. By 1978, investigators from the state's health and environmental departments were scouring the town, sampling milk, water, soil, air, and microwave levels in an attempt to discover why Rutherford was the center of an outbreak of blood-related cancer.

When I visited Rutherford, months after a firestorm of newspaper publicity resulting from the discovery of thirty-two such cancer cases in the small municipality, it was raining heavily. The precipitation had little effect on the air, which still carried a medicinal, perfumelike odor with a rubbery snap, an odor common to industrial smoke and chemical landfills. There was no heavy industry in Rutherford itself, but the town was set on a ridge immediately above one of the most industrialized regions of New Jersey, eight road miles west of New York City. Within a three-mile radius of the town's center were forty plants manufacturing chemicals or allied products, not to mention extensive stretches of wetlands that had been used for landfills.

Mrs. Cleffi had begun to suspect that these environmental conditions might also have accounted for her son Jimmy's disease. "At the hospital in New York, I kept running into leukemia cases from New Jersey—one, two, three, I guess it was seven from our area, close by, in Carlstadt, Kearny, Wallington," she recounted in her strongly marked New Jersey accent. One of the families, the VanWinkles, lived only a block away. "Right after Mrs. VanWinkle's son, we heard of another child who had moved away and had leukemia, and we found an eight-year-old girl in 1977. Then our baby-sitter, who took care of the kids when we took Jimmy to New York, she got leukemia. Then there were two children in the hospital from the area. Then a fifteen-year-old with Hodgkin's disease [a blood-related cancer that affects the lymph nodes]. That seemed odd to me. I could never feel it was coincidence. I first went to our PTA; then we brought it to the board of education and the county PTA. They called a press conference and took it to the state."

Elementary education needs in Rutherford were met by one parochial and five public schools. The largest was the Pierrepont Elementary School, which served an average of 541 students yearly between 1973 and 1978. During those years, and against odds of 10 million to 1, five children in the school district developed leukemia, along with one other child in another section of town and a number of adults. Over all, the town's leukemia rate did not significantly exceed that for all of New Jersey, which, in its turn, had an incidence in line with national rates. But the clustering of cases in an eleven-block neighborhood, coupled with the high rate of Hodgkin's disease found in all age groups in town, had led state officials to suspect that something from the environment was zeroing in on the area, that, in the words of one state scientist, "the kids got zapped by something."

In an unsuccessful attempt to unravel the mystery, state technicians converged on Pierrepont School. They set up electronic gear in backyards bordering on the schoolyard to track any air-borne hydrocarbons, and installed hat-shaped aluminum filtering devices to draw in particulate matter as part of a long-range monitoring program. For more immediate readings, the researchers sucked up air samples in high-volume collectors and large syringes. From these instant tests, technicians were able to detect, in the air and water, no single chemical in sufficient quantities to have caused such an effect, and at the state capital in Trenton I was told that they did not expect to be able to pinpoint a causal factor. In the air, small quantities of benzene, a known leukemogen, were repeatedly detected, accompanied by several other compounds suspected of causing cancer. Seven organic chemicals, including four suspect carcinogens, were present in the tapwater—but not at high levels. "The environmental profile failed to reveal any past environmental condition or event which could be directly related to the clusters," said a DEP report. "Although several sources of pollution, including industry, motor vehicle traffic, and landfills, are located in the vicinity, nothing was revealed which would differentiate Rutherford from other communities in New Jersey's industrialized areas."

Nothing tangible, that is. The state could not dismiss the possibility that a number of harmful chemicals, pouring out in tons from the neighboring smokestacks, had blended together in malign compounds, or that at some time or times in the past a toxic cloud had vaporized from a landfill and wafted into the neighborhood, taking its fatal toll but soon passing away. Chemical contaminants including benzene had been tracked in the soil samples, at low levels near the schoolyard, and there had been frequent reports from the neighbors of an appearance of fallout that looked like green dust.

Pierrepont Elementary School, in the midst of a neat, ornate middle-income neighborhood, is built of worn yellow bricks and has a small fenced playground. Inside, I found myself in a central corridor with pale blue walls and polished floors that led to the principal's office. When I introduced myself as a journalist, a secretary at the main desk became stonily silent. The others in the room glanced up in an uninviting way. After a wait of several minutes, I was informed that the principal would not speak with me. It seemed that the school district's official opinion had already been publicly expressed by the superintendent, Luke A. Sarsfield, who called the cluster of cases a mere "statistical aberration." Nothing more was to be said.

I left the building and trudged across the street into the neighborhood itself, where the first woman I met, Carol Froehlich, happened to be a reading teacher at Pierrepont who had just returned home for lunch. She explained to me that she had seen me waiting in the office but dared not speak of what she heard me questioning the secretary about. She said, "The teachers are trying to keep this thing calm, so it won't affect the children. The first time it was on television, there was all this fear. Parents were keeping their children home from school. There were television cameras all over, disrupting the students. The reporters asked the students if they were healthy, how they felt. The school was unfairly the focus of attention." Despite her criticism of the fashion in which the cluster had been exposed, Mrs. Froehlich expressed to me a deep concern about childhood cancer and told me she was an active member of an antipollution group, We Who Care, which had been formed by Mrs. Cleffi and many other worried parents.

A short way down the street, I walked up a long brick sidewalk to the VanWinkle house and rang the doorbell. Mrs. VanWinkle, whose twelve-year-old son, Wes, had succumbed to leukemia, graciously invited me into a large, tastefully furnished living room beautifully appointed with models of old sailing ships. She sat on a chair next to me and began speaking in a confident voice that showed not the least trace of psychological pain. She too was discouraged with the news media and, unlike Mrs. Cleffi, was not yet outwardly bitter toward industry nor worried about odors in the air. "Until I know who to blame, I'm not going to blame anyone," she said. "I have healthy plants, healthy pets, the other children are fine. And they had the same teachers and classrooms Wes did. I think there was definitely a problem in town, but the media made a shambles of it—false reports in the newspapers and sensational headlines. Who knows if it was susceptibility, something inside, genetic. I'm not ready to blame anyone yet, no one thing. I'm not ready to blame industry." She paused a moment, and in the same firm voice, but this time slightly subdued, she said, "But if I find it's any one factory—if I find it's any one factory, then I'll go over there and I'll tear it down, brick by brick."

Less than a ten-minute ride from Pierrepont School, I entered the rough backroads of a 19,000-acre marsh, tidal creek, and river course famous as the Hackensack Meadowlands. The remnant of a glacial lakebed, it forms a valley of sorts between the Palisades and the Watchung Mountains. Ditches and creeks secluded by cord grass and bulrush crisscrossed the marshland like pretzels, the brackish overflow from Newark Bay. Rabbits and other small mammals live among the tall reeds; the ponds contain thousands of muskrats, waterfowl, and wintering birds of prey. The New Jersey Sports and Exposition Stadium in which the football Giants play looms in the background like a huge tankship at bay. It is not the only development that has subtracted land from the meadowlands. Some 8,000 acres have been claimed by various industries and businesses, an intrusion that began in 1873 when Isaac Singer placed a large sewing machine factory in the Elizabeth Meadows. Warehouses and distribution centers have crept in from the edges, and Newark Airport, afloat on man-made sand drains, gives the swamp an almost urban cast.

Not all of the altered marsh consists of buildings and parking lots. Perhaps most disruptive are the giant landfills that encroach on the tidal garden and the smaller dumps where illegal haulers, oblivious of the implications of their acts, have for years illicitly unloaded tens of thousands of waste drums. Each year in New Jersey, 350,000 gallons of toxic slush simply disappear, and because huge volumes are now sent into the ocean, by 1981, as ocean-dumping bans take effect, more wastes will go to illicit handlers. Their legacy already includes multicolored ponds at the meadowlands' outskirts and rusted barrels that sit in the mud at curious angles. Many of the municipal garbage landfills are now closed, but five private landfills still operate in the meadowlands, in addition to a large plot reserved for Bergen County garbage. Every week 128 communities transfer 30,000 tons of refuse, forming landfills up to 60 feet high and 400 acres in size. Chemical drums, heavily mixed in with the garbage, fill the air above the meadowlands not far from Rutherford with a wide range of toxic substances.

Near the border of East Rutherford and the township of Wood-Ridge, an area bordered by food warehouses and processing facilities, is the locale of the single most serious case of meadowlands pollution. In 1972, after the New Jersey Sports and Exposition Authority had been given permission to build the stadium and racetrack next door, as a concession to environmentalists the DEP and Hackensack Meadowlands Development Commission had ordered the authority to restore 130 acres of Berry's Creek, the central channel in this part of the marsh. It was thought that the creek would make an ideal environmental learning center. That turned out to be true, but not in the sense the officials had in mind. When the authority hired a firm to survey the Berry Creek region, it found that mercury had contaminated the soil for as much as three feet below the surface. The metal bled from the ground in shiny, elusive globules, and it seemed to be everywhere.

The mercury contamination was traced to the Wood-Ridge Chemical Company, a defunct company which had operated for years at the headwaters of Berry's Creek and found the back wetlands a most convenient dump for residues from its fungicidal and mercury reclamation processes. About 300 tons of mercury had found its way into the creek, and it was estimated that there was perhaps another 100 tons in an adjacent swampland.

Life would be impossible without metals, yet in excess they can have a profoundly negative impact on the living cells of the body. In the human body, there are at least 700 various enzymes whose function it is to serve as catalysts, or "reaction initiators," in biochemical actions. Of this number, a minimum of 200 depend upon a metal component for their activity; many enzymes would be impotent were it not for a metal attached to their protein portions, allowing for the internal alchemy that dismantles or builds up substances vitally necessary to metabolism. Once inside the body, ions or atoms of metal, some so small 50 million of them would not occupy a linear inch, hover around a cell and then, depending on their size and electrical charge, pass through the membrane or wall into the cell's electrochemical jelly, where they help determine the cell's behavior.

Mercury has an impressive toxicological past. The eightieth element on the periodic table, it can be found nearly everywhere in small quantities: in rocks, in soil, in the air, and these days, unfortunately, in the flesh of fish. This is mainly because it has leached out of landfills employed by paper mills, paint manufacturers, and chloralkali plants or has been directly discharged into rivers. On Minamata Bay, Japan, an acute and massive outbreak of mercury poisoning began in 1952 after those in the small fishing village unknowingly ingested fish contaminated with the pollutants from a plant on the waterfront. Cats ran screeching through the village and then, in a fit of pain-invoked hysteria, drowned themselves in the sea. Crows stumbled awkwardly upon landing or fell directly from the sky and died. Humans at first suffered dizziness, then difficulty in speaking; then their vision became affected (it was likened to staring into a dark tunnel speckled with small lights) as a result of the mercury's attack on the cerebellum and visual cortices of the brain. In the end people went blind and crazed, or became comatose. Up to 1971, 184 cases had been found, with a death toll of 54, which continued to rise. Of these, a minimum of 26 had been born with the disease, the mercury having passed through the mother's placenta. Other cases have occurred in such diverse places as New Mexico and Iraq, in the latter instance filling the hospital corridors of Bagdad with hundreds of crazed victims who, their brains irrevocably damaged, had to be confined in straitjackets while they waited to die. So overwhelming can the effects of minute amounts of mercury be that it is unsafe to eat fish that contains more than a half of a part per million. Because it is a basic element, mercury can last indefinitely in the environment, changing, by action of the earth's colonies of micro-organisms, into methyl mercury, a far more toxic compound.

There was as much mercury in the New Jersey creekbed as was found in the sediments of Minamata Bay, about the same amount one would uncover in a low-grade mercury mine, and detailed observations determined that the depth of contamination, in some spots, was eighteen feet. To discuss the details of the immense and wanton dumping, I paid a visit to Dr. Glenn Paulson, then the state's assistant commissioner of environmental protection, in Trenton. He was a young and smartly dressed man. "It could be an imminent problem," he said. "The only good thing is that the aquifer is salty and no one drinks from it. The biological situation is and has been that the mercury in the marsh ranges from higher than normal to unbelievably high: up to 19 percent of the sediments, tens of thousands of parts per million. We checked pheasants, muskrats, minnows, ducks, and cattails for mercury, and things people eat, like perch, and we measured the total levels and compared them to the only standard for mercury in living things—fish flesh. Generally, what we found is that in both things we eat and things we don't eat, the levels are below the Food and Drug Administration standards. The conclusion we have drawn is that while it is in the muck, for reasons that are not clear it has not yet moved into the food chain. Why in this estuary has the mercury not moved? That is the mystery question." He then explained that the most likely reason it had not migrated was that Berry's Creek was so polluted by other sources that there was not enough oxygen to support the bacteria needed to change the mercury into methyl mercury, a compound that flows more readily through the environment. Unfortunately, the creek was beginning to show signs of life, a comeback that could heighten the problem measurably.

Two major actions had been taken by the state because of the contamination. First, it was attempting to force owners of the former company to pay the $4 million or so needed to seal the dump and cover an area of thirty to forty acres with blacktop or clay to stop water infiltration. A suit had been filed in superior court by the state's attorney general against the Ventron Corporation and the Velsicol Corporation, two companies that at various times had owned Wood-Ridge Chemical. But the court actions threatened to drag on for years, while five pounds of mercury a day continued to seep into the waterway. The state had also begun testing for mercury in the hair of the residents of the area. Dr. Morris Joselow warned that long-term exposure to the air near the dump constituted a potential health threat, since 1970 air tests had shown levels of mercury in excess of the current federal standards for mercury. Aside from the possibility of air pollution, the dump also directly threatened to contaminate homes within two hundred yards of the swamp through water contact. One day it would be shown that trace levels of mercury had somehow filtered into their blood and urine.

Another distressing aspect was that the mercury was contaminating a wetlands important for the breeding of phytoplanckton, tiny organisms that serve as a staple for fish and invertebrate aquatic forms such as insects. An imbalance in the phytoplanckton population, caused by mercury or the leachate of other landfills, could result in destruction or at least contamination of higher forms in the food chain. If seepages were to kill too many of the organisms, the results would be felt for years.

Driving by one of the streets in a zone that had homes not quite up to the standards of Rutherford, I met Louis Furno, a lifetime resident, who directed me to the exact spots of contamination. He informed me that in the early days the company had been known as Berk's Chemicals and that the plant had emitted fumes heavy enough to choke him and his neighbors and to turn a vegetable garden yellow. In fact, I was to learn that as recently as 1970, conditions at the plant were so bad that the company was warned to take measures for abating the emission of mercury in the workplace, changes that Dr. Joselow, who had been involved in two investigations, said were not effected. When I asked if he knew of anyone who had suffered any ill effects from contamination, he led me to the modest home of Onofrio Mendola, who had worked for nine months in the factory beginning in the early 1940s. From reading a newspaper file on Wood-Ridge, I knew that at least one man, Henry Ruoff, had become unable to talk, walk, or speak and had lost some of his teeth and much of his weight as a result of mercury poisoning in the plant. Mendola claimed to me that in 1943 he too became seriously ill with what the plant doctor told him was mercury poisoning. "We wore masks but it did no good," he said. "I had a rash on my stomach and legs, swollen legs, my toenails were black. I was told to eat the white of eggs, the doctor said that, so I ate eight a day. I was out of work a year. For that, all the company offered was $400." At this his wife, a short Italian woman, joined in the conversation. She had a high, excitable voice and kept staring at Onofrio as she stood in the center of the kitchen with her apron on. She said, "They told us with mercury you either die from it or get completely cured. That's what they said. Thank the Lord, he lived."

While Wood-Ridge's mercury was the biggest threat to the meadowlands, and the Rutherford situation the most confusing, the most acute problem anywhere in the Garden State was several miles down the New Jersey Turnpike, past the small eddies and red soil of the outer swamp, past the mountainous garbage landfills near Newark and the Oranges, at a waste disposal facility inaptly registered as "Chemical Control Corporation." Located on South Front Street, an industrial stretch teeming with bulk-storage tanks for gasoline and liquid natural gas, Chemical Control was in the business of hauling hazardous wastes and stacking the containers, which covered an entire yard near the Elizabeth River, four barrels high. The barrels were unlabeled, unsupported, unstable, and unsegregated as to waste type, and they were in a frightening state of deterioration. In the words of one flustered politician, it was "one of the most hazardous spots in the United States"—located only twenty miles from Manhattan.

In 1979, even before court action could be completed against the negligent firm, the state was forced to begin disassembling the facility and its 34,000 drums. Fumes were beginning to leak from the carriers, posing an imminent fire hazard, and the owners were ignoring state directives to correct the situation. Cases of aniline were stacked immediately beneath cases of nitric acid, raising the specter of an uncontrollable firestorm that would perhaps touch off the huge gasoline storage tanks nearby and at the same time rain down plasticizers, solvents, cyanide, pesticides, and various arsenic residues upon a nearby residential neighborhood. Some of the compounds had the potential of exploding merely on contact with the air. "Not only were drums filled with holes, from which darkened stains flowed down the outside of the drums, but also drums had so disintegrated as to allow vegetation to grow on, in, and through them," stated William Cruice, a consultant to the Department of Environmental Protection. To add to the danger, nitroglycerin was found in some of the mismanaged wastes, and it was noted that welding sparks sometimes flew over the area from other operations on the street.

So acute was the threat of explosion and environmental harm that police or other officials stood watch over Chemical Control day and night as drums were analyzed for their contents and slowly hauled away. Around the first of May, one section of homes and apartments was temporarily evacuated after two drums exploded. The barrels were sprayed with fireproof foam. Later, two Elizabeth police cruising past the plant were taken ill and treated at a hospital after breathing fumes that lingered in the air. The federal government and several industries whose wastes had been stored there began helping the state defuse the time bomb.

In the state's legal actions, which included proposals for civil fines and placement of Chemical Control into a receivership, it was alleged that the company was a front for a quick-cash scheme in which the operators were paid by the barrel to treat the wastes but instead simply let the drums pile up. The situation should have come as no surprise. The former president of Chemical Control was William J. Carracino, who had been given a two-to-three-year sentence for illegal waste dumping and also fined $22,000. "Chemical Control Corporation has never had and does not now have any genuine or separate corporate existence," the state charged, "but has been used and exists for the sole purpose of permitting defendants Michael Colleton, Eugene Conlon, and John Albert to transact their business under corporate guise."

The names of the defendants Eugene Conlon and John Albert also appeared on another state civil action brought against a New Brunswick firm, A to Z Chemical Resource Recovery, which, like the Elizabeth collection yard, was believed to be responsible for dangerous conditions. This firm had been organized by Conlon and Albert, who also managed Chemical Control. At A to Z, flammable solvents, wood adhesives, resins, paint sludges, and other potentially toxic compounds were stored in open-topped storage tanks and 55-gallon drums and fiberboard containers that the state, in seeking civil action, described as "leaking, open, broken, unsealed, toppled over, bulged, and otherwise stored in a manner which permits the chemicals and chemical wastes contained in such drums to be discharged. Spills, leaks, and other discharges have now almost covered the entire ground area in the vicinity of the drums." No one knew exactly all that was stored there, only that the contents were seeping toward an aquifer a half-mile from thirty drinking wells, that children were routinely walking through the unfenced and often unsupervised premises, and that the operations had been conducted without proper permits. Moreover, A to Z had directly defied a 1977 cease-and-desist order from the DEP, continuing to operate despite the risk of a $3,000-a-day fine. By March 1979, a superior court judge took DEP's side, ordering removal of the wastes, proper supervision, and a cleanup of the tainted ground. A to Z showed clearly that there was a chemical waste network in New Jersey, hidden among the clutter of other industries, that was flagrantly defying state orders and, in collecting the profits, posing risks to a large and unknowing segment of the population.

While studying Chemical Control and the network of dumping in New Jersey, I was told by a highly reliable source working in conjunction with the attorney general's office that state investigators, upon checking Chemical

Control's phone logs, found that companies affiliated with a man named Anthony Gaess had frequently been on the line and in fact might have been closely associated with Chemical Control's operations. I was told there had been a number of shipments between Chemical Control and Kin-Buc, another waste disposal outfit which was the object of legal action by the state. In its legal papers, the state had named Gaess as one of the Kin-Buc defendants.

Kin-Buc, in Edison, New Jersey, consisted for the main part of a 220-acre landfill that reached heights of 60 feet. According to state inspectors, it was leaching toxic mixtures into tributaries of the Raritan River and, through the fractured shale thought to underlie the dumpsite, was probably contaminating groundwater in a region cited by the United States Army Corps of Engineers as having potential water-shortage problems.

The federal government sought a series of $25 million performance bonds to ensure proper concealment of the site by Kin-Buc. It also wanted additional corrective measures and charged the company with sixty-nine causes of action, most of which carried $10,000 fines. The actions were drawn under firm, existing laws: the Rivers and Harbors Appropriations Act of 1899, the Federal Water Pollution Control Act, the Solid Waste Disposal Act, and Common Law Nuisance. But tracking down those liable, and legally pinning the blame, was an almost impossible task. The landfilling had been operated through a labyrinth of corporate titles and subsidiaries. Gaess was connected to Kin-Buc through Earthline Company, a wholly owned subsidiary of Wastequid and SCA Passaic. Wastequid, in its turn, was owned by Scientific, while SCA Passaic was run by SCA Services, one of the largest waste-disposal conglomerates in the nation. Kin-Buc itself was listed under New Jersey records as a wholly owned subsidiary of Scientific.

Another company, Filcrest Realty, was also listed as owner of a piece of the Kin-Buc site. Filcrest, in its turn, was wholly owned by Scientific. Yet one more portion of the landfill was owned by Inmar Associates, which had the same address as Scientific. For these and other reasons, the job of bringing the matter to court had been time-consuming and frustrating, effects that perhaps were not unintended by those who formed the corporations.

From 1973 to 1976, the Kin-Buc landfill had received 71,049,932 gallons of liquid wastes—oil, sludges, everything from metals such as lead to benzene and PCBs. Wastes were collected from 350 municipal and industrial sources in fourteen states, and the chemicals had been poured on top of the containers of regular refuse and allowed to drip down the sides; in addition, the state charged, liquids had been emptied into insecure trenches and pools, with no measures taken either to neutralize the wastes (which included carcinogens and volatile compounds) or to guard against contamination of the underground. Noxious fumes had been released into the air by the mixing of certain chemicals, and air samples in a residential area less than a mile away showed levels of polyvinyl chloride compounds exceeding recommended exposure limits. In just a six-month period, Kin-Buc had been cited for no less than forty-two violations by the state DEP for contamination of the Raritan River and its tributaries; in 1973, wastes from the facility were said to have seeped through a meadow, killing a quarter-acre of vegetation.

There had been more serious incidents than that. The state reported that at least twelve substantial fires had occurred at the landfill area, four during an eleven-day period in 1976. They were believed to have been caused by the improper mixing of chemicals. One bulldozer operator had died after a spark ignited a nearby lagoon, and another operator suffered burns when acid squirted upon him from a punctured drum. No one knew what the unholy mix of chemicals and wastes at the dump was brewing.

And the problem was spreading. New York Bay off Jersey City was badly contaminated with mercury and cadmium from its own sewerage authority, and two small waterways, Alcyon Lake in Gloucester and Greenwood Lake near the New York border, had been contaminated by leaching landfills. It was obvious that New Jersey did not have control over its disposal firms and was seriously threatened by the 4.6 million tons of hazardous wastes it generated yearly.

The dangers involved in a large chemical-waste treatment facility were demonstrated on December 8, 1977, in Bridgeport, at a site owned by Rollins Environmental Services. Sparks from a welder's torch touched off an accumulation of chemicals including benzene, toluene, and PCBs, causing a raging fire that sent up a torrent of thick black smoke that resembled a tornado. Six lives were lost in the conflagration. Sergeant John Keller, the first officer to arrive at the scene, told me: "It was just a whole lot of fire, and all the smoke. I had never seen anything like it, and I'm also a volunteer fireman. A lot of firemen got sick. They didn't even know for sure what they were fighting. Pipelines, storage tanks—the whole place seemed like it was on fire. There were cylinders as big as a freight car flying through the air for a couple hundred of yards. Because of the toxic substances, the chemicals, we had to quarantine the four fire companies after, and impound police cars, and put all our clothes in plastic bags. Thirty or forty firemen went to the hospital. The cloud was like a mushroom, with drums popping all over the place, a very black and high funnel, hundreds of feet into the sky."

Dr. Paulson's department immediately dispatched investigators in a helicopter to evaluate the dangers. It was feared that hundreds of poisons were in the cloud, including furans and dioxins, and that they might spread over the area exactly as had the killing toxic cloud in Seveso, Italy, a year before. Since no one in the DEP knew exactly what was burning at Rollins, one of the DEP's scientists had to stand in the midst of the smoke and fumes and, by telephone, read to Paulson the plant's inventory so it could be decided whether to evacuate nearby towns. Fortunately the fire died out within ten hours.

Rollins was eventually allowed to reopen. In response, residents of neighboring Logan Township picketed the plant gates, protesting especially Rollins's application for a permit to incinerate PCBs, up to then a forbidden operation throughout the country. One of the demonstrators, Roberta Quattrochi, commented to the *Philadelphia Inquirer*: "They don't know what they're handling with these chemicals. It's not the Garden State any more. It's the Garbage State. That's what it's getting to be."

8

LOUISIANA: DEVIL'S SWAMP

From the time he was twelve, David Haas Ewell, Jr. wanted his own piece of property. In his family, private land was considered an inalienable right, an extension, in this region of plantations and ranches, of the Southern man. And for a family in Louisiana, it could mean the difference between food on the table and empty shelves. "I saw my father bounce from rented farm to rented farm," Ewell said, when I visited him on his family's large farm north of Baton Rouge. "Had always moved from one cattle-and-hog farm here to one ayonder. He was a hard worker, a *hard* worker, and a hustler, and see, I was the oldest son. He woke me first in the morning. I built the fire. We'd leave at 4:30 A.M. and we'd work the land. He saved all he could and we bought this land in 1940 so we'd have something our own. See, when you own your own property, no one is supposed to kick you off, no one can ruin it for you. That's my idea of what 'property' means."

There were others who felt differently. As the petrochemical industries in Louisiana grew insatiably, they proved devastating to the surrounding farmland, turning creeks into open sewers and permanently disabling the fertile soil. The environmental ravaging here was more blatant than anywhere else except New Jersey, and it appeared to be irreversible, as David Ewell was soon to discover.

The Ewell property, left to the eight brothers and sisters when their father died, was a 1,156-acre ranch of gardens, wooden cattle pens, and cypress trees, covered with green thickets and the shining black of the Angus herd. The barns and living quarters were at the gullet of the gravel entrance road; then there were 600 acres of upland pasture and, in the back, dipping toward the bayous of the Mississippi, a dense and humid mire so foreboding with its plentiful water moccasins and the certainty that a stranger would get lost if he wandered in that it was known throughout the territory as "Devil's Swamp." "It was just a haven for the hunters," Ewell said, leading me and several members of his family to the marsh. "Deer, rabbit, perch, crayfish. They'd go on adown here and make a damn living off'n what they caught. Wasn't much like it. What's it now? Worthless. Nothing left. It's fantastic what they done."

Before we entered the swamp, Ewell, toting a .22-caliber rifle in the event a snake shared our path, swung his red pickup truck into a short driveway just off a corner of his land where, in 1969, a chemical-waste disposal firm, Petro Processors, had located itself. Into earthen pits within several hundred feet of the swamp, Petro poured huge volumes of chemicals and oil sludges from many of the petroleum complexes that dominated the Baton Rouge–New Orleans corridor, tenacles of pipeline rising above the Mississippi's shores. David's sister-in-law, Catherine, who had once spent an entire day keeping tally of the loads delivered, had counted trucks entering the site on an average of one every three minutes. Petro apparently had little knowledge of what it was mixing in its holes. Its invoices and work orders described the debris in vague terms of "clarified oil and catalyst bottoms," or "spent chemical solution." Barrels and viscous sludge went down twenty-five feet in an area bordered only by a levee that, in this exceptionally wet climate, could not have been expected to retain the contents.

We drove to the entrance of Petro, which is now officially closed. I noted only a few pieces of rusting construction equipment in the yard, and a sign that said, "Warning—Possible Hazardous Waste Area." Soon we approached and strode a shaky aluminum fence close to Petro that served as the entrance to the Ewell's rear grazing grounds. We descended a muddy, moss-coated path into the thicket, pushing past the magnolias and hotweed indigenous to the Southern swamps. Although it was July, there was not a mosquito to be seen among the vines, nor were there squirrels or birds scrambling on the branches. David, a stocky, ruddy-faced man who wore a blue construction cap, stopped at the first stream and plunged a branch into its bed, scooping up a blackened clump of noxious sediment with a metallic scent. From the resulting hole flowed a silvery oil, and brown scum floated on the surface of the water. "We tested all this from three to five feet down," he said, "and let me tell you, it was contaminated." With little effort, Ewell broke several brittle branches from one of the dying oaks that lined the murky watercourse and held them close to my face. "You notice how it dies?" he asked. "It's the funniest way of dying. The black stuff runs down the middle. You just smell it. You smell the chemicals in it."

Indeed, the wood had an odor like that of creosote and solvents. But it was difficult to know with certainty if the aromas issued solely from the branch, for the entire swampland gave off a sweet, etherlike odor that rose from the muck, giving me a severe headache.

There was an occasional 55-gallon drum in the bayou water, apparently swept downstream from Petro when the rains were heavy and the Mississippi was high. As I attempted to ford one of the creeks, my right foot slipped and sank into the muddy bank. When I pulled it out, glistening globules of what appeared to be mercury oozed from the hole. In other spots perch floated belly-up, and near one barrel was the shell of a dead loggerhead turtle with a skull about sixteen inches in circumference. Alligators and crayfish had also died, Ewell said. During the Mississippi's annual flooding, the waters in the swamp rose far up the trees and sometimes even over the Petro levees and into the pits, receding to bring the contaminating matter into the bogs and down the bayous. The high-water mark was clearly shown by fragments of junked automobile batteries hanging from the branches of many swamp trees.

None of these phenomena were noticed until Thanksgiving of 1969, when David's brother, E. Q. Ewell, and the sister-in-law, Catherine, who lived in the ranch's main house and actively worked the farm, became concerned over the dozens of cattle that had migrated for the winter into the swamp where the air was warmer and hotweed more plentiful. The hired hand, Dewey Veal, was the first to raise the issue. He saw the animals affected with a mysterious lethargy, their coats stained with what looked like diesel oil. "Yossir, the cattle started to break around the hoof, busted loose around the hoof," Dewey, a black man with gold front teeth, told me. "There was some with burnt hides. The calves had burns near their mouths. I figured, it was from sucking their mothers. Yossir, they was rail thin." Initially skeptical about Dewey's strange tales, the Ewells finally ventured into the swamp to see for themselves. The creeks were blackened with a tarry substance, and the herd was indeed dying. Before too long, 149 cows and calves had succumbed, the hides shrunk tight around their emaciated carcasses. Rains had washed out the Petro levee, Catherine said, and it leaked for months after that.

For years after the depressing realization that more than 540 acres of land had been poisoned, David Ewell attempted unsuccessfully to get help from the government and from a local university in finding out exactly what was in the soil. All were afraid, he speculated, of reprisals from an industry that now dominated the state's economy. "All the state said was, it was okay, we could put our cattle back there," he said. "We didn't know. Nobody knew what was ayonder." He himself paid to have soil samples analyzed and found that in reality the soil had been poisoned with chlorinated hydrocarbons, trichloroethane, tetrachloroethane, carbon tetrachloride, hexachloro-1,3-butadiene, and other nonbiodegradable materials of great and lasting toxicity to the human organism. They were trapped in the soil at concentrations ranging from 180 to 25,000 parts per million.

For the next ten years, digging deeply into a lifetime of savings, David Ewell fought in the courts for the restoration of his land. His brothers and sisters had settled for a total of $90,000 in damages, but he had refused his share, demanding instead that the property be scraped clean and refilled, restored *en natura*. To accomplish that, it was figured, would cost approximately $190 million, with one hundred truck drivers operating continually for seven years to remove and transport the 10 million yards of allegedly contaminated soil and place it in concrete vaults. Ewell sued not only Petro but also those firms his attorneys claimed to have sent waste there: Dow Chemical, Foster Grant, Copolymer Rubber and Chemical, Exxon Chemical, Shell Chemical, Rubicon Chemical, Allied Chemical, and Ethyl Corporation. The civil trial, lasting eighteen full days with the jury excused more than fifty times while lawyers argued over admissible evidence, was the longest such proceeding in East Baton Rouge Parish history, concluding in 1975 with all the defendants found liable. But there was one hitch and a major one at that: the judge would not permit damages in excess of the fair market value of the land. So Ewell was awarded only $25,000 for his share of the land, at $375 an acre, and $5,000 for mental anguish. This hardly compensated him for the expenses of the legal battle—he claimed that in time and money the tab had reached more than $200,000—nor, of course, for the restoration of the property, despite an article in the Louisiana Civil Code that states, "Every act whatever of man that causes damage to another obliges him by whose fault it happened to repair it." Said Ewell bitterly, "I think Petro was nothing but a money-making scheme. Just make money. Never a thought to the people. They're ruining land that can be very productive. A few bastards, dammit, ruin it for the millions. They must've known what they were doing. Now, here I won't eat anything off this place. Those chemicals go right into the fruit—you know, fallout. In the morning there's a dust over the property, so you know it's getting into the grass. And there's got to be a reaction. I just think it's all so uncontrolled, a plaything for the politicians. The governor already said he'd give this in return for jobs. But hell, what good is a job without your health?'

Ewell's concern about a chemical fallout was not an idle one. Beginning in 1972, it was discovered that a 100-square-mile area of pasture lands around Darrow and Geismar, Louisiana, was contaminated with hexachlorobenzene, produced by the volatilization of Vulcan Materials Corporation wastes that had been dumped into local pits. A by-product in the manufacture of carbon tetrachloride and perchloroethylene, the compound was best known for the havoc it created during the 1950s in Turkey, where hexachlorobenzene-treated feed grain was accidentally distributed by the government and 5,000 people were poisoned, reacting with liver deterioration, skin blistering, uncontrolled hair growth, and ultimately, tremors, convulsions, and death. During a routine sampling of beef fat by the United States Department of Agriculture's Meat and Poultry Inspection Program, 1.5 parts per million of the compound were found in the meat of a steer belonging to W. I. Duplessis of Darrow. Soil and vegetation were likewise contaminated, as apparently were people living near the pits, for they were found to have high plasma hexachlorobenzene levels. The dumps were covered with plastic and dirt so there would be no more evaporation. Based on the USDA's standards at the time, more than 40 percent of cattle quarantined after the finding would have been unacceptable for marketing. But the acceptable levels were raised after a USDA petition was sent to the appropriate authorities and the EPA. This change, as well as the special diets fed the animals, which lowered the levels of the poison, averted the destruction of more than 30,000 cattle. No one could be sure how many other animals grazing near dumpsites elsewhere made it to the dinner table undetected. The Ewell land was the recipient, the family charged, of smoke and vapors that emanated from the pit area and burned their lungs and eyes. Ewell suspected that one of the fumes might be phosgene, a gas that in the air, at a mere 50 parts per million, can be rapidly fatal after a short exposure, causing degenerative changes in the nerves and setting free hydrochloric acid in the bronchioles and alveoli; this can lead to pneumonia, pulmonary edema, or lung abscesses.

As if what had happened to the swamp was not discouraging enough, on another side of the Ewell property a second waste disposal firm, owned by the conglomerate

Rollins Environmental Services, had also been landfilling with chemical drums for several years. Catherine Ewell took me back to this part of the ranch and showed me dead or blackened vegetation where she said chemicals had spilled from an open pit. The odor was far more rancid than that in the swamp, and when I looked into one of the pits, I could see stacks of crushed and deteriorating drums, with a thick collection of leachate on the floor of the hole. Nearby, a small drainage trench leading into the swamp had a purplish-red color and a strong odor. Similar smells had wafted near the central house when I first arrived, and Catherine told me that several weeks previously the family had had bouts of nausea and headaches. It was startling to think that a large and once isolated plantation should so suddenly find itself crushed between two obnoxious and ill-kept dumpsites. "That place back there," David said, pointing to a stretch near Rollins. "That place back there had some of the best springs. I used to love to drink that water. See that ditch down at Rollins? There was a *big* spring there. We used to put some cypress boards across there and just stick our heads in there and drink until you just about exploded."

As we walked I noted a pile of cattle bones—the possible result, they told me, of the animals wandering close to a spill. "What's the use of owning land?" David asked. "If they can do this, why own property? Why pay taxes? This land is no good at all. Ruined. When we go down to the swamp, you saw, we stick around in the ground and black stuff oozes out. It's like a nightmare, only the nightmare is real and it's our own land and it's destroyed for all time."

The bad dreams were spreading. Because of the reverence in which the petrochemical industry was held in the

South, plants spewing a tarrish sludge were allowed to operate beyond the regulation of local and state governments. Louisiana and Texas had embarked upon a defilement that one day would surely surpass that of any of the Northeastern states. Like the hurricanes so familiar to the Gulf of Mexico, the destructive elements were building slowly but in great volume, with portents of nothing short of a toxic deluge.

In Louisiana, massive and wanton dumping often had the practical imprimatur of state and parish leaders. Not only did Louisiana produce its own share of hazardous waste—by some accounts the fourth largest quantity in the nation—but it also proved to be a haven for out-of-state haulers, who emptied their tanks into pits, on the ground, to the winds, or down channels that fed into major waterways. (One official publicly estimated that out-of-state wastes outstripped the local variety 20 to 1. More likely, the figure was less than that, but it was still enough to create what some newspapers reported to be 40,000 places where industrial materials of one sort or another were emptied or stored, many on important wetlands that drained into the Mississippi.) In no other state were byproducts handled more sloppily.

That questionable waste disposers enjoyed the endorsement of the Louisiana government was confirmed by some of the curious official actions. For example, in June 1979 the state sliced $402,000, which would have supported environmental attorneys pursuing pollution violators, from the general appropriations bill, leaving only one full-time lawyer to deal with offenders. Nearly a year before, in Lake Charles, Judge Earl Veron, of Louisiana's Western District, struck down federal water pollution controls on hazardous chemicals, ruling that the guidelines exceeded the intent of Congress. The state's loquacious governor, Edwin Edwards, proudly admitted to the media:

"We've made tradeoffs, accommodations, compromises, if you will. Need for jobs, industrial development, and stimulation for our economy justify temporary tradeoffs, in some instances some serious tradeoffs. We did what we thought was best for the people and the economy of Louisiana." The state was "knowingly and advisedly" running the way it was, he added. (Coincidentally, a brother of Edwards's was the legal representative for a waste disposal company attempting to secure a permit for land a mile south of the town of Livingston.)

In one instance, allegedly, government records and drill logs concerning a deep-well injection facility for hazardous materials mysteriously disappeared. During 1978, a Lake Charles man, Michael S. Tritico, director of a volunteer group of environmentalists protesting deep wells at a Browning-Ferris Industries Chemical Services site near Willow Springs, publicly charged that he had seen the drill logs in the Office of Conservation—logs that supposedly showed the porous nature of the underlying rock. A spokesman for the office answered that the records probably never existed. Another alleged case of missing files occurred near Bayou Sorrel in 1978. It was at this site that a nineteen-year-old truck driver, Kirtley Jackson, was killed while pouring material from his vehicle into an open pit. His actions had created a reaction that apparently engulfed him in hydrogen sulfide fumes, paralyzing his lungs. Federal and state investigators were thwarted at first in their attempts to review records of just what had been brought into the site, since the pits had been operated without proper permits. The investigators sought to enter the site itself, and when they were refused, they requested a state court to grant them a search warrant. They were refused that, too, but eventually they obtained a federal warrant. Upon nearing the pits they were met by a distressing sight: the pitch-black cesspools were surrounded by a river, bayou, canal, and fishing lake, and the high-water marks on the trees were at or above the tops of the waste pits. Meanwhile, the local townsfolk had grown so upset at governmental inability to do anything about the mess that they peremptorily stopped further hauling to the holes by burning a bridge that led to the property. After threats of a violent outburst, Governor Edwards declared a six-month moratorium on the issuance of new disposal permits, and months later, an environmental "state of emergency" was declared in the state, though with little real impact on the dumping. "There is big money in these disposal operations," one sheriff's detective, Ralph Stassi, told the Baton Rouge *Morning Advocate*. "You can bet there will be big payoffs."

In Texas, since landfills are huge and unlined, and public awareness of what is being dumped around them is minimal, there have been a number of chemical ecological atrocities, though not yet on the Louisiana scale. Houston and Galveston, the foci of the oil-flare stacks that changed the night sky to orange, were the sites of the worst offenses, but others were found near Dallas and San Antonio, and in many rural and swampy spots. Millions of gallons of wastes had been pumped directly into the most fragile of all ground containments—sand. As a result, percolation into the groundwater or runoff to a river was far more likely than in most Northern states. By 1968, there were at least thirty-three toxic dumpsites in the Houston area. Some of the landfills, I was told, stretch on for dozens of acres, spreading even to swamplands where waste migration is ready and constant. "Texans are only now becoming concerned about solid-waste disposal," said Doris

Ebner, environmental manager for the Houston-Galveston Area Council. "What will happen is that there will be some disaster to make a flash."

Already, some sparks had been seen. The Brazos and San Jacinto rivers were threatened with toxic tinctures. Stenches hung over some of the highways. In 1968, a McAdoo man died while using an acetylene torch to open an improperly disposed barrel of parathion. And then there were the wells. In an outskirts area of Houston, twenty-six domestic wells had to be temporarily closed when some of the 70 million gallons of highly acidic waste poured into a sand pit at Crosby seeped through the underlying supply of water. In addition, failure of a dike at the pit on several occasions had allowed contamination to enter the San Jacinto. Authorities found themselves strapped in red tape when they attempted to deal with the pollution. On January 16, 1968, courts temporarily enjoined the company, French Limited, from open-burning of waste and disposal of liquids, but according to EPA records, operations had continued. On May 28, 1970, violations of a court order resulted in $2,000 in fines. Still operations continued. And they persisted through other governmental actions in 1971 until, finally, a temporary restraining order prohibiting further disposal halted the operation, a full five years after residents first complained of rancid odors. In 1973, in lieu of additional cash payments to the county, the company agreed to deed the twenty-two-acre site to the state of Texas. That appeared to be anything but a bargain for the state taxpayers. What they were to receive was described, to the *Texas Monthly*, by the Harris County pollution control director: "They took just about anything you'd care to mention. Heavy metals, like cadmium, acids, chlorinated hydrocarbons. When you approached the place you had the distinct feeling of descending into the netherworld. From quite a distance you could detect a smell so nauseating you had to wonder how anyone could live near it, and some people did live within a mile. When you got closer to the site, there were lagoons filled with a thick, black liquid. There were yellow and orange flames from ground flares: they were open-burning some of the stuff." In other instances, toxic chemicals had been sprayed on dirt roads or flushed from tank cars and into the ground and sewers.

More serious, perhaps, were the noxious fumes that emanated from a site adjacent to a trailer park near Texas City. Barrels of styrene tar were dumped into one of several pits there, emitting vinyl chloride, a proven human carcinogen. In 1977, a Houston firm reportedly received the styrene waste for recycling, planning to employ it as a creosote extender in the treating of wood railroad ties, unaware that it would be sending into the air a substance that was known to have caused liver cancer among rubber industry workmen. Fortunately the danger was revealed before residents near the plant incurred long-term exposure. Vinyl chloride was also a problem in the growing Houston suburb of Friendswood, where the material had collected at a site once operated by the community's mayor and was later hauled to an unfit construction site. Near other dumps in Texas, adjacent to construction for housing projects and a college, birds swooping down to puddles of what appeared to be water but was in fact chemical leachate died instantly and by the dozens.

The Texas oil industry, which accounted for a third of the nation's production, was spreading around enough noxious brine to transform several areas into dead seas. Three miles southwest of Frankel City, there was a desolate salt lake more than 250 acres in size, where brine mixed with oil and grease had been transported from hundreds of wells at a rate of 50,000 barrels a week. By 1975, there were at least twenty-one known cases of groundwater contamination in Texas, including oil and wastewater in Ellis County, phenols and hydrocarbons in Howard County, and sewage and waste products dumped into abandoned well bores throughout the region. If there was any consolation for Texans, it was that much of the state's wastes were shipped out of state. This could be of little comfort, however, for those areas, like Criner, Oklahoma, which inherited the Texas barrels and the attendant polluted creeks, foul odors, and reports of lead poisoning. (Oklahoma also accepted toxic wastes from Arizona, Arkansas, and Kansas, thereby allowing its air to be infiltrated by hydrocarbons.)

Industries in Texas, like those everywhere, showed minimal concern over where their waste went after it left the back gate, often leaving it to haulers like Jack Arsenault, who at various times operated firms with such names as Industrial Waste Materials Management, B & W Services, and Rabar Enterprises. However professional the titles, operating practices were, to say the least, shabby. According to a presentation at a 1975 meeting of the Texas Water Quality Board, the following was the Arsenault *modus operandi*:

> On January 6, 1975, Texas Water Quality Board inspector Sam Pole contacted Mr. Arsenault in regard to an unauthorized industrial waste site located near Briggs in Burnet County, Texas. Mr. Pole had been told by one of the owners of the land that Mr. Arsenault had been hired to remove all industrial waste from the site and dispose of them at an authorized facility. Mr. Arsenault confirmed that the owner of the site had, by prior agreement, made arrangements with Rabar Enterprises for the removal of 727 industrial waste containers stored at the Briggs location. Mr. Arsenault also stated that he had, in fact, removed all of the waste containers from the location and had delivered them to an authorized solid-

waste disposal site in Purcell, Oklahoma. It was later to be determined that this was false information. On February 5, 1975, the Austin/Travis County Health Department contacted the Texas Water Quality Board, District 3, by telephone and informed them of the existence of containers of industrial wastes stockpiled at a site located at 10959 Research Boulevard. A joint Texas Water Quality Board-Austin/Travis County Health Department inspection was conducted on February 6, 1975. That investigation confirmed the stockpiling of approximately 1,500 containers of industrial waste. Further investigation confirmed that Mr. Arsenault (dba) Rabar Enterprises had been using the location for a stockpiling storage site for at least three months prior to the discovery of the site. Waste materials stockpiled at the site had been generated by industries located in various parts of the state, but mainly in Houston, Austin, and Dallas. Some of the containers had ruptured and many were in a poor state of repair and exposed to the elements. Additionally, the site is located within the recharge zone of an aquifer which supplies water for domestic and stock watering purposes. [When asked if he had any other sites,] Mr. Arsenault stated that the Research Boulevard site was the only location where he stored waste and that he had no other sites. However, he did say that he could have misplaced one or two drums, but he did not think so. Those "one or two" drums were later discovered at a second and larger storage site located in the northwest part of Travis County located near Martin Hill. At this location more than 3,000 containers of waste had been stockpiled, and were basically in the same condition as those at Research Boulevard.

While Texas had better waste regulations than most states, there was a dearth of control over on-site plant disposal, where a large portion of the mishaps occurred, and it was extremely difficult to track barrels across such a large territory to enforce the laws. With few substantial waste collection firms and with the increase in the use of drums as a result of the phasing out of ocean dumping, the plains and floodlands, the dry brown dirt of the desertlike regions, were in clear danger of persistent insult, and would be for years to come.

Among the white-face bulls and muscadines, Louisiana's David Ewell still dwells. He is determined that his fight will go to the United States Supreme Court if necessary, but with each passing month, with each additional legal fee, his chances of victory dwindle. His spirit, unrestrained in its quick wit and dry humor, nonetheless has begun to show the signs of strain and the bitterness of defeat. Perhaps someday he will be able to restore Devil's Swamp. More likely he will not. "I don't want the money," he said. "I want my land. I just can't believe the world has come down to the point that a man can ruin your land and then say, 'Here, your land's not worth anything anyway. Take this.' Somebody's got to make these people understand they just can't destroy a man's property and the environment like this and then cover it up with a little money . . . This fight, this is for everyone. Maybe this is my legacy, to fight this sort of thing. It's for my children. But I must say, at the beginning a lawyer told me to forget it, 'You can't fight industry,' and I must say I'm beginning to believe it."

9

CALIFORNIA: CATCHING UP FAST

One thousand feet above sea level, in a canyon at the side of the Jurupa Mountains near Riverside, California, there are about twenty pits that hold industrial effluents from Los Angeles, Orange County, and Riverside itself. They were constructed on different levels, rising steplike up the granite and sand walls of Pyrite Canyon. This territory is a fascinating one for the mineralogist and the geologist, for it was once a part of the sea's floor. Tremendous heat and pressure had built up against the limestone at the bottom, but the rock mantle held against the strain as they ascended, and instead of erupting into volcanoes, the heat and chemicals created minerals such as blue calcites and stringhamite. Now it was whipped by other forces: the dry winds from the Santa Ana Mountains, and the fumes and washoff of chemical waste.

None of the latter was much to the liking of Ruth Kirkby, a feisty and persistent woman who lived a short distance from the waste site called the Stringfellow Quarry. During the day she could be found on an eighty-acre expanse of property where carobs, ice plants, and salt bush were grown, where peacocks freely roamed, and where a small building displayed a sticker on its window: "Pollution Kills." The Jurupa Mountains Cultural Center, as it was called, was about a half-mile from the pits, and Mrs. Kirkby, an expert on paleobotany and other earth sciences, had founded it in order to teach children and adults about nature's mysterious processes. She was sixtyish, with red hair and a wide face crossed by a pair of gold-rimmed glasses—an authority, she proudly told me, on the fructification of gymnosperms and other arcane topics. As we talked amidst a collection of petrified tree stumps and fossil dinosaur tracks, it became obvious that she was also an expert on the Stringfellow waste pits, which it seems she had been complaining about for sixteen years.

"Here is a canyon," Mrs. Kirkby said, drawing me a map. "Here it is twelve hundred feet; here it is a thousand. They just went in there and made pie-pan pits, and the trucks would come in there and discharge. The Riverside County supervisor, Paul Anderson—his campaign manager, Earl Nutt, was the operator. They located at the site in 1955. It did not even have a dam then. They discharged a few million gallons. Then they kept increasing and increasing. Well, in 1963 my husband and I came back from Colorado with friends of ours. We had driven late, and when we got in we went right to bed. It was hot. We opened the windows. Well, we all choked, couldn't breathe. I slammed down the windows and called the sheriff. People would wander out in their nighties and wonder where the smells were coming from. In 1969, we saw filthy material coming down the site, down Pyrite. It was just half a block from the school! I drove up to the site with a friend. One of the lower pits had broken, and there was a gaping hole as tall as I am. It was right down near the ditch. There, poisons were washing down the street. And in May of 1972 the air was so bad that I would lay awake and say, 'Am I going to have to move? I can't breathe!' I called the air pollution district. They sent a man in fifteen minutes and he came back with pictures. He said there was enough pollution for a thousand cars an hour."

These incidents, documented in papers and news articles Mrs. Kirkby showed me, were the beginning of a furious battle, which escalated in 1972 when she was confined to her home after a knee operation. Sitting in a posture chair next to the telephone, her boundless energy unaffected, she called local politicians to demand that something be done about the Stringfellow site, and organized a group of residents who called themselves Mothers of Glen Avon, the name of the suburban town. Later, the group was to be incorporated into Parents of Jurupa, with its own lawyer. They went to the Santa Ana Regional Water Quality Board—which supported the waste pits and which had as a member Paul Anderson's wife, Ruth Bratten Anderson—to copy every document there that pertained to the pits. Then they went to the planning board and the land-use department and the highway department, doing the same. Dozens of letters were sent to local and state officials, or to the congressman and governor. Fliers and hand-mimeographed newsletters were distributed up and down the blocks. The small group knocked on hundreds of doors in the Glen Avon area to gather signatures on petitions calling for closure of the pits, which were receiving large quantities of sulfuric, nitric, and hydrochloric acids as well as mixtures of heavy metals, organic solvents, and pesticides. "We put out one flier and called it 'Acidgate,' " Ruth recalled. "We began a first-class battle. Mr. [James] Stringfellow's attorney immediately took action. He wrote a letter saying better be careful not to libel yourself . . . Every effort that was made by the community was ridiculed by all public agencies."

The state and county had not even properly tested the soils around the site and were apparently loath to do anything at all. But so great was the public pressure that at the end of November 1972, Stringfellow announced that he was closing the pits after having received about 32 million gallons of waste at the twenty-acre depository. Officials were dismayed: there were few other places for the effluent to go, and without another waste site, plants might have to be closed down and workers relieved of their duties. According to newspaper reports, and records Mrs. Kirkby showed me, General Electric, Aero-General, Philco-Ford, Hughes Aircraft, Sunkist Growers, and other companies at one time or another had sent their wastes to Stringfellow. From then on there was a running battle between the residents and certain segments of the water quality board, which wanted to reverse the decision to revoke the pits' permits. But public pressure was maintained, and the controversy lingered for many years—not only over future use of the pits, but over what should be done with the chemicals already deposited. William Mattox, president of the West Riverside Businessmen's Association, wrote in a letter to local and state politicians: "While we have asked that the pits be eliminated we have been getting nothing but the double-shuffle. It is very cruel for a representative of the people to profess that the acid pits are shut down, when in fact the condition now existing is more dangerous than anything we had from either pit operations or its attendance after operations ceased."

There were many debates over who should pay for remedial action, with the state forced to expend $203,600 from 1975 to 1978 to maintain the deteriorating condition of the pits. Stringfellow claimed an inability to help with the costs. "The corporation doesn't have any money, it doesn't have any assets," he was to explain. "It's got real estate, but nobody wants that. That leaves everyone holding the bag. So we go back and say, 'Well, who was wrong?' I don't feel like I was. I took a good beating and then I said, 'Enough.' "

Public hearings were held on the matter, often turning

into vitriolic verbal matches between Mrs. Kirkby and the officials. She was a dogged fighter, armed with boxes full of records and documents and her penchant for scientific detail. One of her main concerns was the drinking well for the 670 students at Glen Avon School. Through her research, she found that the well had been tested for the toxic metal hexavalent chromium—and proven to be positive. While chromium is essential to the human body, to prevent hardening of the arteries, in excessive amounts it can result in deep lesions on the skin. Taken internally, it can be expected to cause more serious problems, and indeed is suspected as a cause of stomach and larynx cancer. The hexavalent compounds, moreover, are more toxic than other varieties. In Riverside the students, still physically in their developmental years, were regularly taking in small quantities of the compound, and while the levels were far from a fatal dose, no one could assess the cumulative effects. "They got their results one May and didn't retest until the next December," Mrs. Kirkby said. "Later, at a public hearing, they said it was the only time they had found it and it was a mistake. We presented a test showing another well. So I don't believe any of them any more. I've learned this is a sham. They have not based their statements on facts. Since, the school is on district water. But we had to clamor and clamor to get that changed."

At one meeting, at the time the Kirkby group was demanding a clay cover over the site or other corrective measures, Mrs. Anderson of the water quality board, the most vocal proponent of reopening the pits, said, "If you're worried about the school, then move it. It'll be a lot cheaper than paying $500,000 to put a clay cover." She later claimed that Mrs. Kirkby raised the entire issue only because she wanted to expand her cultural center and the pits were in her way. Mrs. Anderson told me she once called Mrs. Kirkby an "evil woman" for "taking economic gain out of their [the community's] good night's sleep." The pits, she said, would be a "good thing" for Glen Avon.

In these years of jostling, I was told by several people, Mrs. Kirkby had been subjected to threats from persons unnamed. For a time she was followed home by a small pickup truck. One morning there was a bullet hole in the cultural center's office window, one of the secretaries told me. Mrs. Kirkby declined to discuss such events in detail, saying, "I just ignored it. I figure you just have to go on living. If you don't do anything, you're a mute conspirator."

As the politicians argued with residents over the lagoons, and among themselves over who should pay for a remedy, the site continued to collect rainfall from the winding canyon and its tributary ravines, and it continued to deteriorate. The rains funneled in from the crevices between the cacti and coastal sage brush as if into a gutter, and after filling the top level of lagoons, cascaded in sequence to those below, threatening finally to smash the twelve-foot retaining dam at the bottom and thereby release an 8-million-gallon torrent. In the first week of March 1978, National Guardsmen were rushed to the scene to sandbag the base of the pits. Two of them were splashed by the liquids and had to be treated in a hospital for burns and nausea. The pits continued their upward rise. To avoid a total collapse of the dam, the director of water quality ordered, at 4 A.M. on March 6, that 600,000 gallons be released down the gully and into the community to take some of the pressure off. A few days later, another 275,000 gallons were discharged. A ribbon of liquid resembling root beer coursed down the mountainside into the flood-control channels under Freeway 60, through pastures where farmers were forced to stand sentry over their herds to stop them from wandering hoof-high in the wastes, then on past Spanish-style yards and into the streets and drainage areas near the school, fanning out into a brownish foam that soaked into the land which, soon after, was to become the location for housing developments. Children and animals cavorted in the discharge, and adults waded through the puddles to and from their cars. One report said that the effluent's chromium content was twenty to eight hundred times federal drinking water standards, and surely there were many other harmful chemicals present as well. Other tests had identified DDT, nickel, cadmium, lead, chloroform, trichloroethylene, and other potential carcinogens at the pits—bad news indeed for those who drank from the Santa Ana River, into which the flood channels flowed. Nor was it encouraging for those using their own wells. Mrs. Kirkby claimed that the substances had flooded a front yard and some cellars. A phenolic pungency permeated the canyon.

One resident, Rickie Clark, told the *Los Angeles Times*: "The main thing that always goes through my head is, Can my kids have normal kids? My kids played in that stuff when they were pumping. That's what got me involved. I didn't know. I sent them down there to play. We had just moved here, and the other kids would go down there to play. It's pretty there, you know? Nobody knew what it was. They never told us. But then one day they came home and told us there were suds in the water. I thought somebody was washing their car. And one of my kids came home and her boots fell apart after she played in that stuff. The soles came off at the bottoms and I had to throw them away."

As a substitute for a full inquiry and a complete remedy, California authorities began pumping out the liquids to a landfill in West Covina and covering the pits with clay. The contaminated dirt would remain underneath, perhaps one day to poison the huge Chino Water Basin.

But if Mrs. Kirby had partly won her battle on toxic wastes, California, like the rest of the newly industrializing Sunbelt, was losing the war. While the smog above Los Angeles was the most visible pollutant, fires were erupting at landfills, containers of pesticides were finding their way into unsecured sites, and oil brines, no longer pumped entirely to the sea, were being indiscriminately dumped into the ground. The state that prided itself on its heightened consciousness, ecological and otherwise, was often quite conventional in its dangerous disposal of the more than 100,000 tons of hazardous wastes it produced each month, 92 percent of which went straight into the ground. It was routine for workers to be injured handling drums of pesticide wastes, or for nauseating fumes to waft from lagoons for days and miles at a time. Despite some innovative regulations that controlled waste handling better than just about any other state, and some well-designed landfills, certain of the state's practices were as primitive as those of the back counties of Georgia and Tennessee. Mounds of municipal garbage were being piled in the canyons, and chemical effluents were pumped directly into sandy ground. Municipal garbage was mixed with toxic wastes in a fashion that would have been considered remarkable even in New Jersey. "Currently, in the eastern portion of Los Angeles County, they are using waste generated from that county literally as a sponge to soak up toxic industrial wastes and then utilizing burial and compaction techniques," complained a Riverside health official, Dr. Jerrold L. Wheaton, in a letter to Congressman George E. Brown. "Why are we collecting the most toxic by-products industry produces in locations now known as dumpsites that have the potential of becoming 'environmental time-bombs'?"

The problem was an immediate one. Contra Costa and other counties had so much hazardous waste traveling their roads that no agency could be expected to keep track of it all, even though the state did have a manifest system that supposedly kept records of disposed chemicals. The evidence? San Joaquin Valley: alluvial deposits contaminated by salt and lye from the olive-processing industry. Fresno and Tulare: boron-rich waste water from citrus packagers seeping through the ground. Montebello: aluminum and acids mixed in a waste tank-truck exploded on the road, closing the Freeway there for more than an hour. Belmont: a tank of left-behind waste fumigants was mistaken for propane, and seventeen people were hospitalized from the fumes, including firemen whose protective gear they penetrated. Visalia, where pentachlorophenol was moving toward an aquifer; Copperopolis, where waste asbestos was suspected of having entered the environment. Upon testing water wells in twenty-four counties, California authorities found a third of them to contain significant quantities of dibromochloropropane, which is suspected of causing cancer as well as sterility.

Many of the wastes did a good deal of traveling before they found a grave. They were handled by haulers who, despite their professional pretenses, had been known to mix dangerously reactive substances, locate their ponds and landfills on poor subsoil, and allow emission of air pollutants far more toxic than the exhaust from cars.

One of the better firms was Industrial Tank Incorporated. Still, at one time, ITI had facilities in Contra Costa, where residents complained that the fumes were taking the paint off their cars; in Salona County; and in a site known as

"Martinez," where the company received several air-pollution citations and where blue smoke was sometimes seen hovering over a pond. Some of the odors were said to be hydrogen sulfide, a rotten-egg-like emanation that can quickly paralyze the respiratory system and has done so, too often, in chemical plants. In September 1975, ITI was contracted by Shell to dispose of some highly volatile organohalogens. The company determined that the highly toxic wastes were not suited for its own ponds, and so they were trucked to West Covina, to a dump owned by a company called BKK. BKK rejected the materials after an initial inspection (it too had had its share of odor complaints, and had once accidentally emitted cyanide gas). The next try was at the Richmond Sanitary Services site, where the substances were accepted and run into an evaporation pond. Soon a noxious white cloud arose and hovered for several hours, entrained in the wind, until authorities were forced to evacuate several nearby buildings where people had suddenly become nauseated. It was not simply a local problem, however; there were reports that the plume had traveled ten or fifteen miles downwind, before dispersing near Alameda above an unsuspecting public.

Californians had a tendency to use huge amounts of organic phosphate pesticides. Inevitably their residues were found up and down the Pacific coast. Pesticides of the phosphate variety do not linger in the environment or the human body with the same tenacity as the chlorine derivatives. Their threat is in their quick assault upon the body, in which they rapidly head for the nervous system, resulting in tremors, convulsions, and often death. Two of the most common of the organophosphate group, used extensively in California, are the pesticides parathion and malathion. Already in the official state records are hundreds upon hundreds of cases where workmen handling waste drums that contained phosphate insecticides suffered severe injury and death. In 1969, about seventy drums containing ingredients for parathion production were found buried at a regular dump in the city of Cabazon. The discovery was made after thirty-seven of them, unearthed by a long winter's rain, washed down the San Gorgonio River, threatening places downstream such as Palm Springs. Meanwhile the bay near San Francisco was laden with toxicants.

Farther north, in Oregon, are large stockpiles of chemical residuals not only from that state but from surrounding states and Canada as well. In one instance, 30,000 drums of pesticides were buried in the sands of Alkali Lake; and near Arlington deep trenches were bulldozed and dynamited in the rocky plateau so that they could be filled with acids, pesticides, and mercury.

Similar problems are found in the state of Washington, where mine tailing piles, waste settling basins, and landfills dot the countryside. Mercury-tainted grain seed was unprotected at some dumps; residues from TNT production had entered some aquifers and inhibited salmon eggs from hatching. At a Kaiser Aluminum and Chemical Corporation plant in Mead, near Spokane, pollutants were settling into the wet areas of the ground. In Spokane itself, the Department of Social and Health Services found trichloroethane and tetrachloroethylene in the city's water source. Zinc and other metals migrated from tailing piles along the Coeur d'Alene River, which flows into the Spokane. Paint sludges and solvents were poured into Washington's abandoned coal-mine pits; they trickled into aquifers through fissures in the rock. Stored eight years on an open lot in Puyallup, near houses, a school, and grazing cattle, were hundreds of barrels of dioxin-related compounds—distilled residues of orthobenzyl parachlorophenol, a germicide. Some of this material had been sold as a wood preservative, to be brushed upon barns and fences in a most hazardous way, while other quantities seeped through deteriorated containers and into the soil. It was years before officials took any action, or in fact even knew of the dangerous conditions, and though the company from which the residues originated reportedly declared what remained to be "innocuous" and "inert," the Department of Ecology later classified these substances as "extremely hazardous."

Whether from oozing tar pits in Utah, or sugar-mill wastewater in Maui, Hawaii, or chromium-poisoned water in Colorado, where the federal government had also handled large quantities of mercury, PCBs, and DDT, the Western states had become a disaster waiting to happen. Before long there would be a health outbreak somewhere in the large rangelands, or near the deserts, or along the rivers that flowed near the heights of the Rocky Mountains: that was inevitable.

10

MICHIGAN: SOMETHING IN HEMLOCK

Since 1974, Michigan farmers have destroyed more than 30,000 cattle, 150,000 chickens, and millions of eggs in an effort to purge from the state's food supply a chemical known, for short, as PBBs. This compound, a flame retardant, had been accidentally mixed with livestock feed by a company later purchased by the Velsicol Chemical Corporation, causing milk cows fed on it to dry up and sicken by the droves. For many months, state agricultural officials were unable to pinpoint the cause, so that by the time it was identified, enough meat and dairy products had made the rounds to create detectable accumulations of PBBs—polybrominated biphenyls, a suspected carcinogen—in the bodies of 90 percent of those living in the state. In Lansing, there were sighs of relief at the state's various agencies when, a short time ago, the last large load of contaminated cattle was rounded up, killed, and hoisted into a specially designed landfill.

But officials may not have much time to relax. Since then, indications have arisen that health dilemmas may be starting anew in every corner of the Wolverine State. In the western part, drinking wells supplying sixty homes in Muskegon County were condemned because of 10,000 drums left rusting in a warehouse yard by a small chemical firm now defunct. For months it was feared an explosion there would send a lethal cloud into a neighborhood nearby. In Livingston County, at the southern end of the state, a midnight hauler unloaded 2,000 gallons of toxic paint thinners and the defoliants 2,4,5-T and 2,4-D into an open farm ditch fifty feet from a child's swing set and near to a house and pond, leaving the state to clean up the mess. And of course, there were the 2 billion gallons of contaminated groundwater in Montague.

But in these cases, as well as the myriad others plaguing the state, at least the environmental technicians could see or smell the problem. They wish the same had been true for a bizarre medical mystery that was unfolding in a small town just southwest of Saginaw Bay named, appropriately, Hemlock.

That something was askew in the old German farming community of Hemlock first occurred to Carol Jean Kruger. At forty-three, she was a brawny, parch-lipped farmer who lived on one of the largest dairy farms in an area of flatlands known for its cider mills and bluegrass. "In 1974, things kind of started," she said, as we stood near her red brick house with its yellow-pillared porch. "But it took until 1977 for me to put it together. When we did, we could go back to 1969 and see the problem."

It was in 1974 that she bought ten Holstein cows, only to watch during the next six months as two of them rapidly lost weight for unknown reasons and died. Not long afterward, large numbers of calves also succumbed mysteriously—forty in one year alone—and in a grisly fashion. Their teeth were stained lavender and brown. Their hind legs were grossly swollen. Chronic open sores developed on bald areas where the bristly hairs had fallen out in clumps. Although many of the cattle had arrived at the farm boasting a healthy production record of up to 25,000 pounds of milk a year, the flow, once the animals settled down on the Kruger farm, was barely a trickle. To prove her point to agricultural inspectors, who claimed at first that parasites and poor maintenance must have been to blame, Carol Jean began piling the telltale carcasses behind the central barn. The display did little to sway the officials, but for Mrs. Kruger the exercise was an interesting one. She observed that the dogs would not eat the carcasses and that one calf, heaped on the pile in the warmth of late September, lay there for two weeks with no signs of decay. Horses panted, ran high fevers, and died with tails turned brownish-purple; and two Shetland ponies developed hooves that curved upward like the pointed shoes of an elf. Occasionally, said Mrs. Kruger, the odor of "burnt brake linings, or spent carbolics," permeated the milking area, and an oily substance coated small puddles of water outside with an iridescent sheen, leading local humorists to allege that the neighborhood was about to become rich from an oil strike.

Other animals fared no better. Rabbits had large tumors on their ribs and innards. Cats wandered away and disappeared. On the barn floor, in the meantime, several mice were seen dashing in concentric circles and then suddenly collapsing. The eggs of fancy pheasants, peacocks, and chickens failed to hatch. Among the birds that did hatch out, two geese had their wings on backwards. It could have been a coincidental quirk of genes, but to Carol Jean it was unnerving.

Among those neighbors living in the fifty homes nearest her, Mrs. Kruger said she counted 22 cases of bone and joint problems, 17 of kidney and bladder difficulties, and 16 of lumps and tumors. Skin rashes and teeth "that crumbled like tissue paper" were also prevalent, she said. She herself had a brownish tincture to the whites of her eyes and her toenails, and lost weight even though she was eating more than usual. The top row of her teeth broke off at the gum line.

After months of complaints, state and county authorities sent a medical questionnaire to a sample group of one hundred homes. The survey did not include physical examinations and therefore did not provide objective proof as to the cause of the disturbances, but officials said it uncovered more complaints of rashes, dizziness, urine sugar, visual problems, limb numbness, and other maladies than in a control group chosen in the town of Blumfield. "It is apparent from the analysis that, for health problems as a whole, sex and age do not explain the differences between Blumfield and [the Hemlock] area," said state health director Dr. Maurice S. Reizen, without venturing a theory about possible causes.

Because the problems were common to both animals and humans, Mrs. Kruger came to her own conclusion on the common denominator: life around her, she speculated, was being slowly and secretly poisoned from the water that was pumped from their wells.

Virtually all of the thousand or more people in and immediately around Hemlock drew their drinking water from well fields only a short distance from the Kruger farm. For the past several years many of them had noticed a difference in its taste. The water had always been somewhat brackish, but the saltiness seemed to be increasing all the time, and once in a while it tasted a bit like cleaning fluid. In their sinks and bathtubs there were darkening stains where the water most frequently came in contact with the porcelain. For these reasons, Mrs. Kruger began distilling her well water and passing it through charcoal. She believed that as a result her complexion, which had turned yellowish and scaly, was markedly freshened. She recommended the same process to her ailing neighbors, who soon experienced fewer headaches and fainting spells and felt less nausea.

I drove up Pretzer Road past unengaging fields of sugar beets, oats, and navy beans that stretched on for as far as one could see, interrupted by occasional clutters of spartan frame houses and aluminum silos. About a half-mile from Mrs. Kruger's house, I passed a small cemetery plot near Saint John's Church, and turning onto a dirt road, I found the home of the Jungnitsches, a folksy couple.

Ed Jungnitsch, a thin man with an unquenchable thirst, was picking up stones from his yard one September day in 1976 when he began to ache all over his body. Returning to the kitchen, he quaffed down two quarts of water. That night, at a gathering with friends, his mood turned uncharacteristically irritable, "like he thought everyone in the world was against him," in his wife Kathryn's words. The following Sunday morning, Ed sprawled out on the living-room sofa and stayed there all day, immobilized by the same splitting headache he had suffered throughout the previous day and by pains and swelling in his extremities. Toward afternoon the middle toe on each foot began to feel numb, and by nightfall, Kathryn had taken him to Saint Luke's Hospital in Saginaw.

It would be fifty days before Ed left the hospital. His speech turned progressively incoherent, his motions awkward and strained, and by the end of the first week he had become totally paralyzed, unable to raise a hand off his lap or cough or swallow. "I couldn't concentrate long enough to say a prayer," he said. Eventually he improved to the point where he could use a walker, but when he returned home and began drinking the water, he sickened again.

Nor was the rest of the household without medical problems. The two children, Jennifer and John, had had bouts of skin rashes, bladder problems, tonsillitis, lethargy, or headaches. At two years of age, John's backbone had still not properly fused together. At the same time, Mrs. Jungnitsch developed a cyst on her ovaries, disorders of the bladder and kidneys, an enlarged spleen, severe headaches, and tingling sensations in her arms and legs which precluded a restful sleep. She had also lost a child. Not long after Ed's strange collapse, Mrs. Jungnitsch found a lump under one of her arms, and during the following six months similar growths developed in her groin and neck and at the base of her skull. Her doctors at first diagnosed her problem as Hodgkin's disease, or cancer of the lymph nodes. Lymph-gland disease is not unusual in Saginaw County, where the mortality rate from Hodgkin's disease in white males is significantly above the national average. But after removing several of her lymph glands the doctors decided her problem was other than cancer. "We talked with Carol Jean and bought a distiller," Mrs. Jungnitsch said. "Ed got better. He's to work now. We all got better. My 'cancer' disappeared. After a while the doctor suspected the water. He told us to boil it. But it didn't need killing. It needed taking out!"

From the distiller, Mrs. Jungnitsch regularly removed a viscous residue and tossed it on one part of her lawn. Soon, no grass would grow there.

There were other unappealing parallels to what had happened on the Kruger farm. Again, animals were the best barometer. Once the Jungnitsches butchered a doe fawn that had white spots covering its liver; their Labrador retriever refused to eat the scrap venison. Down the road, Gary Krischer told of a rabbit he had hunted that had green meat throughout its edible parts and of tapwater that, when spilled on his refrigerator, had taken off the enamel. Another neighbor lost a small herd of cattle to a strange affliction with symptoms similar to those of Mrs. Kruger's animals.

In the Jungnitsches' back lot was a pen with four geese that honked loudly as we approached. One of them appeared to have spikes piercing through the feathers where the wings should be. In actuality they were deformed wings, facing the wrong way. "You think that's something?" asked Mrs. Jungnitsch. "How about the chicks? We had chickens born with their guts on the outside, or like that when we opened the shell. They were bantam chickens, and some geese, too—about forty-four in all. They had their bellies on the outside of their bodies. Looked like hickory nuts—that was the guts."

The county health commissioner, Dr. Senen L. R. Asuan, was not inclined to acknowledge a problem. He saw no need to conduct an extensive survey or perform liver function tests or take thorough medical histories. "We consulted a physician there and he didn't know the answer either," explained Dr. Asuan brusquely. "You can't tell if there is a problem, based on headaches. That's subjective. There is nothing medically tangible there. We are sitting tight and seeing what else is in the area—looking for more reasons to go in and do more of a survey. So far, we don't have any plans."

On February 8, 1978, the Michigan Department of Public Health sampled the Kruger wells, finding a high level of salinity but "no connection between the described health problems and the water supply." It was not that the state was immediately discounting the bizarre reports. The technicians were well aware of the capricious nature of groundwater, how it could fluctuate dramatically from time to time; indeed, in later testing, other state technicians found the conductivity of the water in one well to vary 50 percent over samples taken just prior to and after a bioassay (test on animals), indicating a significant change in the nature of the aquifer.

Still suspicious of the water, and dissatisfied with the

governmental response to their requests for an extensive investigation, Mrs. Kruger and the Jungnitsches took it upon themselves to have their wells tested by an independent laboratory, Raltech Scientific Services of Madison, Wisconsin. This time, laboratory researchers found diethyl ether and extremely low amounts of toluene, freon extractable oil and grease, and trichloroethylene. While the compounds were theoretically capable of producing many ailments, the quantities discovered were not large enough. Nevertheless, an important question remained: What were these solvents and industrial substances doing two hundred feet below the surface, in a farmland aquifer?

Only after the residents informed the state of their analyses did the Department of Natural Resources' water quality experts initiate their own samplings. In some tests the department tracked toxic compounds such as carbon tetrachloride, aromatic amines, PCBs, and phthalates, three of which were federal "priority pollutants," and also an unidentified halogenated chemical. But again, they were at trace levels, and secondary tests failed even to confirm their presence. "It's something that could be missed, I suppose," admitted Robert Courchaine, director of the DNR's water quality division. "You may not find it today and it may have been there yesterday or last year. But right now, we have no reason to believe there is widespread organic contamination of the aquifer." Chloride and sodium levels were high, but not to the point where they were thought able to cause such damage. Iron and manganese concentrations exceeded the recommended levels for heavy metals, and in one sampling phenolics were found as high as 7.6 parts per billion, quite in excess of the 1962 recommended limit of 1 part per billion established by the United States Public Health Service. In the three municipal wells the PCB levels were up to a half-part per billion, certainly a level unwelcome in the kitchen sink. Having spent only seven days testing some of the Hemlock wells, the water quality division decided to halt its analyses in order "to spend more time on problems we know about," according to Andrew Hogarth, chief of groundwater compliance, who described Hemlock as a "quandary."

What the state did *not* look for in its review of the situation may one day be as important as what it did search out. No formal study was made of the exact groundwater patterns of the area. No monitoring wells were sunk to check strata just below the wells. No regular surveillance program was instituted to check for sudden variations in contaminant levels. And most important, the technicians failed to analyze well-bottom sediments, where any industrial compounds that had once infiltrated the system might have accumulated.

There were other questions that went unanswered. During its study, the DNR had dropped small crustaceans known as *Daphnia* into containers of wellwater to test for toxic effects. The *Daphnia* could not survive in one sample, and the state, in a final report released in April 1979, admitted that it was at a loss to explain definitely why.

Dotting the entire Hemlock area is a series of industrial wells used through the decades to extract oil and brine from the limestone, shale, and dolomite below. There are at least fourteen such borings in the vicinity of the Kruger farm alone. Most important are the brine wells. It was because of these that in 1897 Herbert Henry Dow took an interest in founding a chemical company fourteen miles to the north, in Midland, where it was to grow into one of the ten largest chemical concerns in the nation. From brine wells in the Hemlock area, the Dow Chemical Company extracted sodium, chlorides, bromine, magnesium, and other ingredients vital to the chemical industry.

Once spent, the brine must be disposed of, and Dow decided that Hemlock would be a logical place for that operation too. Pipelines were installed to carry the salty wastes into reinjection wells, discharging the useless and contaminated brine back into what is known as the Dundee Rock Formation, 2,500 feet or so below the potable water supply. There were at least four reinjection wells along the two roads parallel to the Krugers' street, two of them within a mile of the problem-plagued farm. Into those two, Dow was reported to have pumped between 3 billion and 7 billion gallons of spent brine. The discharges had stopped several years before, except for sporadic usage, it was claimed. One well was being operated in October 1978, but was closed by the state's geological survey when concern arose over the troubled reports from Hemlock.

There was no evidence that Dow's waste wells were contaminating the drinking water. In fact, geologists considered it unlikely that the spent brine, confined below a thick partition of shale, could migrate upward to the point where it would taint the drinking wells. But no one knew if there were fissures in the shale, and the existence of the old oil wells—some of which may not have been properly capped—raised the possibility that waste liquid had risen up through them, bypassing the shale on its way to the surface.

The state of Michigan was not sure exactly what was in the spent brine, or whether other wastes had also been injected. While Dow maintained that nothing more than spent salt water was sent to Hemlock, the state said only that they had "no information about the nature of substances injected previous to 1974 or of the nature of the production brines." There was concern over what else might have found its way below, for Dow's Midland plant manufactured or used as intermediates dozens of highly toxic compounds and produced, in great quantities, herbicides laden with tetra dioxin. State investigators said trichlorophenol waste was pumped underground at Dow's Midland site itself. This was an interesting revelation, for although the site was a substantial distance away, there was still a possibility that fractures might have developed in the earth that could carry the waste for that distance. Such a possibility was raised by the EPA, which cited A. P. "Dutch" Beutel's account in *The Dow Story*, when, speaking of his experiments with deep-injection wells at Midland, he recalled: "What I had done with the pressure was to lift the earth and fracture the zone. The brine wasn't being forced into the sand—it was pouring into the fracture." In the summer of 1979, when the United States Environmental Protection Agency finally began a close review of the situation, one of its staffers noted that fluids containing the herbicidal ingredients had a higher concentration of sodium chloride than of calcium chloride. "Therefore," said an EPA report, "the salt concentrations of the brine line fluids at Midland, unlike the salt concentrations of the brine line fluids in Hemlock, resemble the salt concentrations of the wellwater in Hemlock." Furthermore, dioxin had been detected in fish taken from the Tittabawassee River near Midland and also from the Shiawassee River nine miles south of Hemlock. Because the symptoms of dioxin poisoning so closely resembled some of the town's health complaints, an EPA memorandum concluded that there was "good reason to be concerned about dioxin contamination." Yet up to that point the state had not even tested for dioxin in the well sediments (they said doing so was expensive), nor had the department traced the flow of groundwater from the Midland plant.

There has long been disagreement over the safety of deep-well injection of hazardous wastes. Some governmental regulators and industrialists consider it the most responsible way of handling chemical and perhaps even radioactive garbage, far better than the use of surface landfills and more economical than neutralizing the wastes. Critics decry the method as worse than sweeping debris under a rug—discharging material into a layer of land where no one can keep track of it. There are now perhaps as many as 400 deep-injection waste wells in twenty-two states, and their numbers can be expected to grow vastly in the coming years as industry attempts to limit its use of landfills. While the United States Geological Survey reportedly finds the deep-injection disposal technique acceptable, the states of New York and New Jersey, among others, totally ban its use. In Oklahoma, Louisiana, and Texas, where deep-injection wells are widely employed, controversy has swirled over their safety and their number, which, if oil wastes and shallower wells were included, would probably far exceed the 400 figure. In most cases deep-well injection involves simply the drilling of a deep hole, the insertion of a metal casing halfway down, and then the pumping of waste material into porous layers. There have been some problems. In Dade County, Florida, a shallow well leaked old battery acid into Miami's aquifer. In Colorado, at the Defense Department's Rocky Mountain Arsenal near Denver, as I've mentioned, the pressure of materials in a 12,000-foot-deep injection well caused small earthquakes, and the well had to be capped. In Erie, Pennsylvania, an injection well reportedly backed up to the point where wastes were spewing twenty feet above the surface like a geyser; ultimately, 4 million gallons of wastes were spilled. Not long after the Erie episode, then Interior Secretary Walter J. Hickel described deep-well injection as a potential "Frankenstein monster." Still, the EPA has taken its time in formulating regulations on underground disposal, and during these delays cities such as Chicago and Saint Petersburg, Florida, have spoken of plans to divert wastes underground, planting the seeds of future contamination. The Chicago plan was an $8 billion one in which sewer overflows during heavy rains would be stored and transported in a 132-mile underground tunnel system built 150 to 300 feet below the surface—and not far from the groundwater that supplies the city's suburbs. Elsewhere, oil brines are routinely forced underground in a way that is certain, someday, to cause an environmental dilemma. There is simply no way of knowing for sure how far or in what direction the wastes will move.

Aside from the possibility of upward migration of Dow's wastes, there was also the chance that pipelines transporting the spent materials from Midland had leaked their contents near the surface and polluted the groundwater below. The brine pipelines had been known to leak in the past, and they ran parallel to several of the roads where the loudest health complaints had been voiced. According to the state, there were twenty-four reported losses of brine from the Midland plant's pipelines between June 1977 and November 1978, including one of 24,000 liters. "We do not have complete records as to the numbers, locations, and volumes of brine losses prior to May 1977," said the state report in 1979. "It is possible that significant losses may have occurred during that time period which are not documented." The pipes were buried about three feet below the surface. When the contents of one spill were analyzed, high levels of chloride and sodium were found, along with lesser quantities of freon, extractable oil and grease, phenolics, and fluoride. Next to the wellheads in Midland was a small pond in which the DNR had reported the presence of ammonia, bromide, and organic carbon. In addition, Dow had said ethanol, methanol, and acetone were present in spent brine, at one level or another. Whether such ponds had once contained hazardous wastes and whether these could have seeped into the ground was unknown, according to the state.

Dow sent technical experts to speak with Mrs. Kruger, and they accompanied the state during its 1978 well samplings. The company says the subsequent state study cleared the firm of any blame for the Hemlock problems, and it declined detailed comment on the matter when I queried headquarters. "There is no way we can win in a situation like this," said Dow spokesman Thomas Sinclair. "We've done everything humanly possible. There's nothing there."

In June 1979, Dr. George Waldbott, a Warren, Michigan physician who specializes in environmental health and has written two books on the topic, told me he had examined eight people from Hemlock and believed they were being affected by chemical contamination. His "tentative diagnosis" was that some of the people had been sickened by fluoride while others exhibited symptoms suggestive of PCB, phenolic, or bromide poisoning, or a combination thereof. "I'm convinced that every single one of them had sustained some environmental damage," he said. "But it's hard to hang a hat on one substance. It's speculative. I'm reasonably sure Mrs. Kruger's condition is due to PCBs or

PBBs because of the brown nails and skin pigmentation. Other people's conditions could be related to both fluoride and bromide intoxication." He was sure that some of the animals showed symptoms of fluoride exposure, especially one cow with mottled teeth. But there were also what appeared to be clear-cut cases of fluoride poisoning in the people. "It is difficult to make a diagnosis in retrospect. I would say that I wouldn't be surprised if these people had consumed a good deal of poisoned food."

It was Dr. Waldbott's concern that the fluoride content of the water, while not extraordinarily high in most analyses, might have enhanced the toxicity of other compounds in the water or become itself more toxic because of interaction with these compounds. While the other substances he cited had been tracked at times, and while it appeared that there were PCBs in Mrs. Kruger's blood, the state did not think the levels could have been as damaging as Dr. Waldbott inferred. The physician also said that there seemed to be too many miscarriages and other pregnancy problems in Hemlock, as well as bone disorders he attributed to fluoride, but here too there was a dearth of concrete supporting data.

There were potential ground-contamination sources other than Dow, but no one could determine if they could have migrated to Hemlock. Other plants in the general vicinity had waste-water stabilization and seepage lagoons, one handling PCBs as well as tile-field discharges from another facility. Thirteen miles west, in Gratiot County, was a forty-acre landfill containing massive quantities of PBBs that were showing signs of leaching, and a deep-injection well which was owned by Velsicol.

Yet another possibility was that the people were suffering from toxic particulates in the air. "My main criticism with the state report was that, besides no testing for dioxin, there was no evaluation of the air," commented the DNR's James Truchan. "I have no doubts the health concerns are real, but pinning it down is a very complex thing . . . Hemlock is near an oil refinery in Alma and the Michigan Chemical Company, which had to release bromides because it was manufacturing the PBBs." Others too thought the air a more logical route.

Among those Dr. Waldbott examined was the Wiechec family, who lived near a brine pipeline about a half-mile from Mrs. Kruger's farm. A combination of arthritis and bone growths, skin boils, and other ailments had convinced the physician that something out of the ordinary was occurring. The Wiechecs' property was located along a stretch of fields crisscrossed by swales and drainage ditches that sometimes did not freeze over in the deep of winter. They had a brownish appearance, with an occasional fleck of what looked like oil. The Wiechec lawn, when I visited them in the spring of 1979, was inundated with groundwater.

Whether or not their well was to blame, it was clear that the family had its problems. Irene Wiechec, at forty a plump woman, had lived in Hemlock all her life and said that like her parents and three of her sisters, she had developed diabetes as she grew older. Like her own children, Irene said she found herself plagued with rashes, sore throat, and hard skin boils she called "corebuncles." She had been operated on nine times for foot ailments—bone spurs and malformed toenails—and one daughter had undergone foot surgery seven times. Irene also related a history of six miscarriages and a partial hysterectomy.

The night before I stopped by, Irene said, her husband, Leonard, who worked in a General Motors plant in Flint, had been rushed to the hospital with several frightening symptoms. He had developed a bright red nose that was infested inside with cysts. When she had him soak it in a plastic bowl of water, a red vertical streak suddenly appeared across his forehead and he became incoherent, stuttering and shaking. At the hospital his problem was diagnosed as "staph infection."

Before I asked any more questions, I had Irene bring me a jar of water. In it I observed brownish particles settling to the bottom. It was quite unlike the residue I had seen at the Kruger house; that had been white and syrupy. The Wiechecs had no distiller but had just begun importing bottled water. Many other families, however, continued to draw from their taps.

Sitting with Irene was her son Carl, who, she said, was missing a kidney. It had been removed because it was woefully stunted, the size and hardness of a walnut, with a small growth on one side. The other kidney was beginning to bleed into his urine. The boy had been born partially deaf, she explained when I tried to ask him a question directly.

There were two daughters in the family, Rosemarie and Becky. Rosemarie had missed much school because of kidney and bladder disturbances, I was told; she was embarrassed because her hair occasionally fell out in clumps. But Becky had an even more difficult time with teasing classmates. This I learned when Irene excused herself from her modest kitchen and returned from a back room with something in her cupped hand. She then placed on the palm of my left hand several teeth that had been pulled from Becky's mouth. They were permanent teeth, and all appeared to be deformed. The largest was also the most unsightly: split down the middle, it looked like a pair of rabbit ears, and the enamel was black. "The dentist just goes on shaking his head," she said. "It's something she can't help, and me, I can't help, and it's getting disgusting."

Although I could smell no chemical odors from her tap,

Mrs. Wiechec claimed there had been times when the water smelled "like burnt plastic, or something." Odors also came from the direction of a Dow line near the back property of her home. "I have uppers and lowers, and the water gets on your teeth and makes them slimy," she said. Her parents would have been able to tell me more history of the pipelines, she commented, but both had died from cancer, one stomach, one kidney.

Spurred in July 1979 by a hearing of the Senate Subcommittee on Oversight of Government Management on governmental handling of such cases, and further prodded by the criticism of Michigan Senator Carl Levin and an internal EPA memorandum that cited the situation as a "potential imminent hazard," the EPA region for the Michigan area has begun to look more closely at the Hemlock case, and the state has also shown renewed interest as of this writing. But because of the complexities and the financial resources that would be necessary, and the elusiveness of some of the possible contaminants, it will probably never be known what happened in Hemlock, or rather, what *is* happening there. There are fresh reports of similar illnesses among residents who live two miles west of the Kruger farm in the more populated part of town, including a stretch of several blocks on Sandridge Street where dizziness, limb numbness, skin rashes, and internal disorders have been reported. Perhaps all the cows had been smitten by parasites and the people by a foreign virus. Or perhaps a combination of coincidence, hypochondria, and bad genes had come into play. Or, perhaps, these people were stung by a sudden slug of contaminated groundwater that, during its subterranean caprices, infiltrated their pumps only to disappear soon after, without leaving much of a trace.

The sole certainty is that Michigan and other Midwestern states such as Illinois and Ohio, with their intensive chemical and automobile manufacturing, will probably be facing other Hemlocks in the future. So too can one be certain that, whether or not they were at fault in Michigan, deep-injection wells are bound soon to cause serious problems throughout the nation—and perhaps equally indecipherable ones.

The day I left Hemlock, I went to survey the farmland at 6 A.M., anxiously hoping that, because of a previous night's rainfall, something I could not have seen the day before would be visible. I spotted one sheen of oil, but where it came from could not be told. In the neighborhood was an injection well consisting of a small green aluminum shack with a large pipe running into the ground, not unlike an oil derrick, and accompanied by valves and gauges. Under one pipe was a puddle; it too had an oily look. There were few promising clues.

Amid the guttural groans of the remaining cows on the Kruger farm, as they waited for their morning feed, Carol Jean handed me some of her mail correspondence with politicians and health officials, which spanned a two-year period, and saw me off to my muddied car. She sighed as she stared at her front-yard well. "We just want to find out what's wrong," she said haltingly. "That's all we ask."

11

MARYLAND: THE TRIAL OF DR. CAPURRO

Several miles from Elkton, Maryland, in the northeast section of the state, there are rolling hills covered in maples and oaks and ash trees. On one sloping hollow, just down a steep road and past "Milbourn's Eggs," there is an area known as Little Elk Valley.

Partway up one of the hills sits a small house that since 1943 has belonged to Dorothy Logan and her husband, John, a materials handler for a nearby Chrysler plant. The living room is small, a comfortable room the Logans have paneled and decorated with religious images, among which, on the color television, is a picture of the Virgin Mary. At the opposite side of the room, hung prominently in the middle of the wall, is a striking painting of a little girl four or five years old, with blond hair and a pensive expression, sitting in a wicker chair with a glass of milk in her hand. It is set off by a dark background.

In 1961, in the burned shell of an old paper mill at the bottom of the valley less than a mile from the Logan home, the Galaxy Chemical Company quietly began the business of reclaiming waste solvents from companies along the East Coast. Impurities were removed by distillation, or the components were separated by fractionation and then put into lagoons or pits so the organics could vaporize. Such processes have often been considered desirable by environmental officials, not only because such a process reduces the amount of industrial waste that is buried but also because it theoretically makes it easier for official watchdogs to keep track of where wastes go.

Those who lived near the Galaxy plant did not have such a positive attitude, though. Soon after the company's arrival, the air in the valley alternately took on fragrant, ethereal odors and, on the less agreeable days, the smell of skunk effluent and other unpleasant emanations.

"Oh," said Mrs. Logan, "it was so strong at times. It smelled like everything from a strong tomcat to vomit to glue to gosh-darn rotten eggs. We felt it shouldn't be allowed!"

Once Mrs. Logan was hospitalized for a week because of vomiting, and she believes it may have been from the air pollution that arrived with Galaxy. Doctors told her that much of it was just her nerves, however, and one of them assured her that "nothing that smells like that would bother you." The Logans did not move away, nor did other old-timers in the valley, as a matter both of principle and of finances. They were there first, and they did not think it was fair that they should have to go into debt for another home.

From 1961 to 1976, 685 complaints about the valley's air were made to Cecil County and the state of Maryland. Residents were suffering from headaches, nausea, chest pains, dizziness, and other ailments they had not been plagued with before, some of which precisely corresponded with periods when the vapors were strongest. At night, looking up toward the street lights, residents saw greenish and yellow fog. Illnesses improved when they left their homes for a vacation or a hospital stay. "My daughter Sharon had something like pneumonia," Mrs. Logan drawlingly recounted. "She was given medication for it. Well, so she went to camp, she was gone a week, and when she came home she was fine. Then she got sick again and I had to take her to the hospital."

I asked her about the painting that had confronted me on first entering. She took a deep breath and, with a firmness in her voice, told me that she bought it to remind her of another daughter, Diana. "I tried to protect her," she said. "I kept her in the house, shut up, and even bought an air conditioner. I didn't let her play in the yard. She was so sad. She would put her nose to the window and look out at the yard." Then Mrs. Logan said that at 5:40 A.M. on November 17, 1968, Diana "went to sleep" as she lay in the hospital on a mattress filled with icewater in a futile attempt to bring her temperature down. She was five years old and a victim of acute leukemia. "They told me she would live six months," said her mother. "That was when she was two. But she had good care. The doctor told me I did a good job. She lost her hair a couple times and had some sores, and when they went to give her a transfusion they couldn't find a vein. But there was not much pain at the end, there was no hemorrhaging. I was there. After she died, the doctor came around and told me, he said I had done a good job."

Mrs. Logan does not say for sure whether she believes chemicals are to blame. After all, the child had Down's syndrome and thus was more likely to contract such an illness. Suspicion, though, remained. Mrs. Logan heard of another leukemia case in the neighborhood, and other residents had been hospitalized for the same illnesses as herself.

In 1967, a bizarre and perhaps historic controversy developed over the small, sloppily controlled waste plant. That year, an Italian-born pathologist, Dr. Pietro U. Capurro, who directed the laboratory at Union Hospital seven miles away, moved into the valley's general vicinity, about a mile and a half downstream from the plant on the Little Elk Creek, in search of what his wife described as their "Shangri-La." Their house was a two-story brick type, dug into the side of one of the hills, with a brick lane lined with azaleas and holly. They did not know about Galaxy's problems when they first moved in. They were unaware of the plant's corroded pipes and bad pumps and of the chemicals that seeped from the plant and into a creek, their fumes snaking with the waterway as if attracted by a magnet. "It was July 4, 1968," Mrs. Capurro said. "We had the windows open and all of a sudden—streams and streams and streams of odors were coming into the house. You don't know how hot and humid it was. The whole valley reeked."

Even before that—in fact, their first weekend in the house—Dr. Capurro had smelled what he knew to be solvents and had begun inquiring about the source. Several weeks later, he called Paul Mraz, vice-president of Galaxy, to complain that the air smelled like benzene, toluene, and xylene. There was reason to believe that the doctor had an unusual ability to distinguish such vapors. He had once been overcome with fumes in a hospital laboratory and left temporarily paralyzed, and henceforth remained especially sensitive to airborne solvents. Besides the experience, Capurro had studied at Mount Sinai Hospital under Dr. Harry Goldblatt, who had done pioneering work on chemical carcinogenesis in animals.

At about 11 P.M. one evening, Mraz stopped by the Capurro home and denied that such solvents could be coming from his plant. When Capurro challenged him, Mraz, who had initially entered with friendly gestures, grew irate. "You can't even speak English," he told Capurro, whose English syntax was often choppy and incorrect.

Mraz, however, had badly underestimated the pathologist's determination and outrage. Obsessed with proving his point, Capurro began to keep logs of various odors, wind patterns, plant conditions, and chemical levels in the valley air. He conducted analyses with a gas chromatograph, and sometimes with mass or infrared spectrometry, upon samples sucked into an air-tight syringe or scooped from the creek into screw-top containers. By the time this investigation ended about eight years later, Capurro had discovered a situation that seemed more acute than the Love Canal calamity, if not as extensive.

In the air of Little Elk Valley two compounds were present, benzene and carbon tetrachloride, that are indisputably recognized as human carcinogens. Other harmful substances—toluene, methyl ethyl ketone, tetrachloroethane, methylene chloride, and at least nine other compounds—were also detected by Dr. Capurro. Often awakened by odors in his house, or by abdominal and chest pains, the doctor would scurry about with his equipment, taking readings inside his own home or near others in the valley. On one occasion, Capurro found carbon tetrachloride *inside* his house at 90 parts per million, higher than the levels in some of the most troubled sections of 99th Street in Niagara Falls. Usually at times of high pollution, the air levels inside the homes were about a third of what they were outside, but there were other occasions when, trapped by closed windows and the right atmospheric conditions, the levels inside were higher than in the outside air. Levels seemed to increase along the creek in the lower regions of the valley, while at points perhaps only fifty feet away the concentrations were minute.

Starting in 1970, Capurro, aided by the Thiokol Chemical Company and the Maryland Health Department, began continuous monitoring from two units. One was mobile, in the trunk of a car, while the other was stationed about a mile to the south of the plant. From these readings, Capurro discovered that the vapor currents followed not only the creek but also roadways and other contours and man-made structures, moving in uneven streams and eddies instead of diffusing evenly like water poured into a bowl. Because of this phenomenon, certain areas were afflicted far more than others by the mostly invisible, sluggish fumes. It was obvious that the creek was badly polluted. Samples taken by the state at Capurro's insistence showed what appeared to be ethyl acetate, amyl alcohol, butyl alcohol, toluene, and trichloroethylene. A black, viscous material taken fifty yards downstream of the plant showed a heavy concentration of toluene (3.3 percent) in the air immediately above it. So it is logical to assume that chemicals were vaporizing from the tainted water and contributing to air problems for those living nearby on the banks. Capurro told me that a neighbor's child became ill after swimming in what was once known as a good trout stream.

The stream's pollution, according to a myriad of reports, came from spills on the Galaxy property. One worker was quoted by a reporter as saying that he had personally seen at least fifteen spills go into the creek. Those were the result, he said, of an operator not properly watching a distillation unit; consequently, vessels would overflow and the mess would be pushed toward the creek. Another source of possible leakage was the lagoons of sludges and slurries adjacent to the Little Elk Creek. Curious as to just what was kept on the property, Capurro slipped onto the grounds one night to collect a sample from a lagoon and fell into the waste pit. Mrs. Capurro told me that when he returned home, they wrapped his shoes and ruined suit in a plastic bag and threw them away. "The hair came off his legs!" she said. "And when some of it got splattered on the bathroom door, the paint came off! That was at the beginning."

Some of the solvents detected in the air and water, including the carcinogens benzene and carbon tetrachloride, were also tracked in residents' *blood.* Capurro reportedly had discovered nine solvents in blood samples, toluene and tetrachloroethane among them. In July 1969, one patient was found by Capurro to have a level of hippuric acid in his urine that indicated toxic-level exposure to toluene. That same summer, Thomas Evans, who lived near the plant and was later to suffer from pancreatitis, had his blood tested by a chemist for the DuPont Company; graph peaks of benzene, toluene, and methyl ethyl ketone were present. In other cases, it was noted that such blood levels decreased proportionately to the length of time a person was away from the valley.

In the light of those ominous findings, it was only natural that Capurro began to tabulate various complaints of illness in the same way he was detailing environmental contaminants. In June 1970, Capurro reported in the journal *Clinical Toxicology* that of 43 people who had been exposed to the solvents, all were suffering from chronic fatigue (especially in the pockets of high pollution), 36 had acute burning-throat episodes, 33 had subacute irritability, 32 had headache problems, 28 complained of burning eyes or light-headedness or both, and 24 told of chest pains, which usually traveled underneath the rib cage, radiating to the left side and back along the tenth rib. Of 8 persons with such pains, 7 had an elevated lipase level, indicating problems with the pancreas, liver, or other digestive organs. This study was done in 1969, the year of the most complaints.

What was particularly frustrating about some of the symptoms was their nebulous nature, which caused doctors to dismiss many of them as psychosomatic. Nondrinkers in the valley described transient feelings of drunkenness when passing the plant, and chronic feelings of indecisiveness. But a chemical link was impossible to prove.

Nonetheless, anecdotal evidence was convincing. Capurro reported instances where two or more people collapsed on the same morning, when the air was bad. An eleven-year-old girl was admitted to the hospital one day with the chief complaints of headache, nausea, and vomiting. She was found to have an abnormal glucose tolerance, and allegedly, an analysis of her blood indicated benzene and acetone. Upon her removal from the area, she showed a sudden three-hour increase in blood sugar. Yet another case involved an eight-year-old girl who showed tender liver, keratitis, headache symptoms, nausea, and a swollen face. Capurro reported carbon tetrachloride in her blood. Her brother also had once been admitted to the hospital for pains in the upper gradient region and for swelling of the face. That such effects would crop up through sheer coincidence seemed unlikely.

Most prevalent among the illnesses were those affecting the pancreas. Pancreatitis, in fact, seemed to occur in nearly epidemic proportions. Such illnesses were accompanied by elevated lipase and abnormal glucose tolerance. Illnesses of the liver, kidneys, and lungs were also pronounced.

Capurro's gloomiest moments must have come in the early 1970s, when he began counting victims of cancer in the valley. The figures vary, but it has been reported that Capurro at one point judged leukemia to have occurred fifty times more frequently than chance should account for, malignant lymphomas fifty-eight times more often, and pancreatic cancer at similarly high rates, indicating the existence of an extraordinary cancer cluster. Over all, Capurro and an associate at the hospital laboratory, John E. Eldridge, computed the cancer death rate in the polluted areas to be seven times that of Cecil County as a whole and more than five times the nation's average. Capurro's "conservative" estimate, at one point in the study, was that of 15 deaths in the valley among 120 people between 1967 and 1973, 8 were due to cancer. Later, some of his figures were to rise dramatically.

When a special task force composed of state and county researchers followed up on Capurro's loud and startling claims, they too found an inexplicably high cancer rate. While their figures were not as high as Capurro's—there were differences in the size of the area surveyed and in the length of residency for those reviewed—an interim report by the task force stated that lymphatic cancer deaths were between ten and a hundred times those expected, depending on the boundaries of the study group. Ten cancer deaths were found in the plant area, and all the victims had lived there five or more years. Peritoneal and prostate cancers were statistically significant, as were cancers of the lymph system and organs associated with red-blood-cell production. "There is an indication of both spatial and temporal cluster of lymphatic cancer deaths in Little Elk Valley," the first report said. "All four cases occurred within a population of several hundred and within a period of six years." Clearly, something extraordinary was happening near Elkton, Maryland.

At hindsight, both Capurro's study and that conducted by the state leave something to be desired. Capurro's record keeping, while understandable in one who was conducting the epidemiology in his spare time, was not as detailed as a scientist might want, and there were questions about his selection of population boundaries. In turn, the state admitted that a complete medical evaluation of all present and former residents, plus inclusion of a control group, was beyond its capacity.

But the state examination did present individual cases in more detail and pointed to interesting trends beyond the existence of the general cancer pox. On one table was listed the death in 1975 of a sixty-eight-year-old "housewife" from cancer of the pancreas. As an addendum, it was mentioned that her husband had died the following year from "malignant lymphomas of abdomen and chest" while her son, at forty-five, had succumbed to leukemia. The bare state account described the George Feehly family, who had run a small general store about a tenth of a mile from the plant's west side, overlooking the waste site from another rolling hill. They were known as feisty, colorful people, and like the Logans, they would not move away. Reportedly, in an argument with Mraz over the odors, Olive Feehly was told that if she did not like it she could move. "You couldn't even work in your garden," Mrs. Feehly told a reporter just before her death. "Something like that has no business moving in here and injuring our health and making it impossible to go outside. Chemicals aren't to be mixed with people." She insisted that she would stay and Mraz would be the one forced out. Olive also complained that since Galaxy moved in, she was spending most of her time visiting funeral parlors, a remark that too quickly twisted into morbid irony.

Despite the inundation of evidence that something very unusual was occurring in the valley, the county and state never showed interest in supplying enough manpower and equipment to break the case open. In 1971, the plant was forced to close for a short time until it could upgrade its filtration of emissions, but the official watchdogs had little of Capurro's initiative and persistence. The complaints by Capurro and other residents often were not followed up immediately, so that by the time an official arrived, the odors had dissipated. Even when some of Capurro's most disheartening findings were presented to them, medical and environmental officials scoffed or procrastinated, in no hurry to confront uncomfortable realities and face up to a remedy. One of Capurro's attorneys told me that government agencies tended to treat the doctor as a "crackpot," interpreting his obsessive interest and learned alarm for fanaticism, while, at the same time, Capurro grew all the more determined and vocal. "I asked them to follow the people and see what happens in the future. They show little interest," Capurro said. "Is very bad. A disgrace. Ridiculous. Even the EPA!" He also complained about Carl York, an investigator for the Maryland Bureau of Air Quality. In a 1975 letter to Dr. Emery Lazar of EPA, Capurro charged York with what he fairly or unfairly perceived to be too cozy a relationship with Mraz. Asked Capurro in the letter: "Are they [the health authorities] corrupt or simply incompetent?"

Similarly, there was not much cooperation from other doctors. Capurro sensed that they were afraid of being dragged into the courtroom by Mraz, and they too quickly dismissed problems as psychosomatic or pegged other incorrect diagnoses. An article on Capurro in *Today's Health* related the account of one mother whose children were waking up during the night and complaining of pains in their arms and legs. The family physician said that the ailment was simply "growing pains." Three valley residents were treated as epileptics, not as victims of unhealthy fumes. And a ten-year-old boy who complained of leg pains and vomiting was told to soak in Epsom salts. "One of the children was taken to a nationally renowned medical center, and no diagnosis was reached," Capurro wrote in an article for *Medical Tribune*. "All these physicians failed to include one important factor in their examinations—the environment. Universities may be remote from the problem. It is difficult for the overworked physician to pay attention to this field. An additional difficulty is that tests of air and water are required, and the cost of these often becomes a problem. Universities and medical centers should train health workers to study environment. Then at least they would know when to suspect environmentally caused illness and when to call on air or water pollution specialists for help. At present, this kind of teaching may be affected by political influences in some schools. Many of the best trained scientists are employed by industry."

Throughout his lonely battle against pollution, Capurro's family underwent an inordinate amount of stress and harassment from sources unknown. Their mailbox was destroyed seven times. A tire on their blue Camaro was slashed. Nauseating chemicals were thrown on the front lawn. One evening in 1970, Capurro was near the Feehlys', looking down at what he describes as an ongoing Galaxy spill, when suddenly, he alleged, a station wagon bolted out from the plant and chased him off the road. "It came so close!" he recalled. "I jump—between two cars. Oh, very close. In 1969, I told my family to go to Italy, it was getting too hot. When we talked to the police, they say they can't do anything. But they taught us to shoot. So my son, five or six, learned to shoot." The doctor bought a rifle, and he admits to once firing out into the valley darkness when a rustling was heard near the house. During the controversy, there were claims that Capurro had threatened Mraz's life, and Mraz, according to some of those I spoke with, told people he was afraid of what Capurro might do. A corresponding fear was obviously felt by the doctor. One night the Capurro family heard shots outside their house, and the next day Capurro noticed what might have been a gunshot scratch on the brick. No one ever linked all those problems with Galaxy, but it remains a fact that two stubborn men had locked into a long-standing and bitter feud over the plant. Even after he moved out of his house in 1972, Capurro says the harassment continued, including vandalism of some of his records and household belongings.

The dispute continued in diverse Maryland courtrooms. The first legal salvo came from Capurro, who along with several other residents sued Galaxy for causing ill-health effects and won $34,932. No compensation was awarded for possible permanent damages. Next, it was Mraz's turn. In June 1975, Galaxy and its principal officers sued the doctor for $2.5 million, claiming "injurious falsehood, disparagement of property, libel, slander, and defamation." The trial is widely described as a most unusual, almost eerie battle, or in the words of Capurro's attorney, Richard Magid, a "traveling circus" that set a record for the longest trial—three weeks—in nearby Caroline County. "There was a mixture of everything," said Magid. "You have to understand the flavor of this. Caroline County is probably the most rural place in the state. And here they're listening to this unbelievably complicated case, a case that is probably unprecedented. Judge Thomas Evergnam, he would speak to the jurors, all farmers, about their cows. 'So how many are you milking this year' and that kind of thing. You also have to understand Mraz and Capurro. They had been going after each other for years. Mraz will sue at the drop of a hat. The publicity hit him hard. To this day, I don't think he believes he created a problem. Then there were the expert witnesses—incredible. I don't think I'll have anything like it again, ever."

Well-known scientists such as Dr. Samuel Epstein from Case Western Reserve University, a cancer expert, testified for both sides of the debate. Epstein was in Capurro's corner, but he was neutralized by witnesses that Mraz's attorney brought in, and both sides kept the pace brisk—so rapid, I was told, that Mraz's attorney at one point keeled over with a slight heart attack. Capurro's big problems, according to his attorney, were his strong personality and the fact that his studies were not models of scientific procedure; the pathologist was somewhat disorganized and anecdotal in his approach. Along with that, his selected study groups were such that his final figures, representing those who lived closest to the plant and in the most troubled areas, produced inordinately high anomalies unrepresentative of a large population. With such a tiny population, a chance rise in cancer can be interpreted as a frightening epidemic when in fact it is not. During the trial, questions abounded concerning the latent period of cancer, which is usually estimated at about twenty years but which some scientists now argue can develop much faster. Could Galaxy have caused such a problem so soon? And even if the cancer rate was high, was Galaxy the cause? While Capurro never directly stated in print that the plant was the cause, his inferences were strong enough to leave a reader in absolutely no doubt of what he was talking about. As the doctor once said about the plant, "If you're in a room where everyone is scratched and there's a lion in there too, then you've got to suspect the lion." The plaintiffs argued that a paper mill once located at the same site—a mill that handled dyes—could have been to blame, although that plant was closed long before Galaxy opened its shop.

Years before the trial even began, physicians throughout the East had been polarized by the issue. Dr. Jesse Roth, then chief of the diabetes section at the National

Institute of Arthritis, Metabolism, and Digestive Diseases, said that the valley residents "are getting a real screwing. Whatever is going into the air is doing a hell of a lot of damage." In opposition were people like Dr. Franz Goldstein, a Philadelphia gastroenterologist, who reviewed health records at Union Hospital and disputed several of Capurro's diagnoses, and Dr. Joseph G. Lanzi, an Elkton general practitioner who had treated some of the valley residents as far back as 1959 and felt some of the illnesses preceded Galaxy and were "bellyaches" as opposed to pancreatitis. In a 1972 article carried by *Medical World News*, Dr. Lanz was quoted as saying, "The people [in the valley] have quite a bit less than high school educations, and when a man with a supposedly great knowledge of medicine [Capurro] walks into their homes wearing a gas mask, holding his belly, and saying he's been poisoned, naturally they're going to develop symptoms." By the time the trial of Dr. Capurro began years later, the same sort of questions still abounded, with the addition of claims by Mraz's side that Capurro had committed outright fraud during his diagnoses. It was hard to deny that Capurro was by nature an excitable and single-minded person, not a controlled, objective scientific performer. But the last word remained on Capurro's side when, on the eve of the trial, the state released its study showing a high cancer rate near Little Elk Creek, and a state epidemiologist, Dr. Barry Friedlander, who was later to describe Capurro to me as a "man who did a great public service," took the stand in one of the few low-key moments of the trial and systematically presented evidence that buttressed if not fully supported Capurro's claims. The verdict: a unanimous decision in Capurro's favor, just before Christmas of 1977.

But the real test for Capurro was to see what remedial action would be mandated. That result was far less satisfying than the court trial.

Little Elk Valley today remains the same rural niche that, with its nicely varied flora and calm country roads, might still attract uninformed urban refugees seeking a healthy environment. Children ride horses to the polluted creek, and in the background, from the same hill where Mrs. Logan's house sits, comes the lowing of cows and a chicken's occasional chatter. Galaxy itself is small and appears almost abandoned, a mottled brick building in the back, a tall smokestack, and large tanks and vessels along with the numerous waste drums. The entrance—unnamed, unpaved—is blackened with an oily soot, within a child's stone's-throw of the first homes to the east. Near the rear of the plant, where I stood overlooking the barrels stacked near Little Elk Creek, the odor of solvents was distinctly perceptible. I saw children making blackened snowballs to the north of the site, down the road from Feehly's vacated, boarded-up store. Many of the homes at the base of the valley show signs of poverty—broken appliances in the front yards, splintered porch steps, old cars that will never run again. I walked with Lula Schwemmer and Martha Jackson in front of one of the run-down homes; Martha told me about the headaches she sometimes suffers, while Lula said that fumes still come from the plant in the early morning hours. Although virtually all the long-time residents said the situation has improved, Mrs. Logan confirmed that there are still bad days once in a while but didn't want to elaborate because of the fear of getting sued. She and the others have given up fighting.

Away from the valley, Mraz, who has now finished law school, and who would not speak to me, is still going after Capurro, suing him over another situation concerning a gravel-and-stone quarry that Galaxy used as a disposal site when it was pressured into deactivating its on-site lagoons. That new dumping site is located miles from the Galaxy plant, near Route 40 just outside Elkton proper. It had allegedly created another nuisance situation and provoked a family, the McMillans, into suing the small firm for health damages. While that court attempt was not successful, it irked Mraz into countersuing not only the McMillans but also Capurro and a second medical researcher, who Mraz apparently felt were urging the McMillan suit against Galaxy and providing inaccurate medical data. Galaxy was forced to clean up the quarry chemicals. It was reported that wastes from the plant are now taken out of state to an undisclosed location.

When I last spoke with Dr. Capurro, whose contract with Union Hospital had not been renewed, he was in San Gabriel, California, recovering from what he said was pancreatitis and a nerve disorder that may be related to exposure to solvents. Twice thus far, he claimed, he has been offered hospital positions only to be subsequently turned down "after certain phone calls were made." He also mentioned that his medical license in Maryland was not being renewed. Whether or not that had anything to do with his involvement with chemical contamination—and there was a good chance that it did not—it seemed somewhat odd in light of his distinguished background, which includes instructorships at Albert Einstein College of Medicine and the University of Illinois and a post as director of the laboratory at a Saint Louis hospital. Moreover, he was once described by an authority in the field as one of the best pathologists in the East. What he would like to do is continue work on the toxic effects of chemicals, but he seems psychologically fatigued and filled with anxiety. While he still churns out an occasional medical-journal research paper on the subject, despite the constant threat of suit, he is not being compensated for toxicology research, nor are other physicians who have taken up the task. "I am in limbo," he complained in a distressed voice. "I want to work. Maybe in Elkton. Maybe I will go back to Elkton and fight, do you think? From California I cannot fight. They will sue me everywhere. Maybe I go back to Elkton. Right now I am so confused. I have been harassed, terribly, and it's so confusing."

It can be said that both Capurro and Mraz won and that both also lost. The physician succeeded in bringing about a precedent-setting investigation, but in the end there is Galaxy, still in business. Capurro is the one who was cast away.

III

LOSING CONTROL

12

MERCHANTS OF POISON

The awesome difficulty of controlling pollutants is implied in the sheer tonnage of chemical production. Whereas 25 million gallons of the solvent benzene were manufactured in 1940, by the early 1980s the figure will approach 2 billion gallons in the United States alone. In the last fifteen years the production of general organic solvents rose more than 700 percent, while during the same period plastics shot up 2,000 percent and synthetic fibers more than double that figure. One way to prevent the by-products of these chemicals from falling into the hands of the bootleg dumper would be to make central points of waste collection available in each region, or perhaps in each state. There residues could be analyzed and recorded. They could be recycled or treated, properly packaged, and then carefully and safely discarded.

In theory this scenario is appealing—experts in the trade handy to mop up the toxic mess. Yet in reality, precisely such a waste industry has already developed throughout the nation, and, as seen in Elkton, Maryland, at best the results have been mixed. Accumulating unprecedented quantities, waste disposal companies find themselves in a quandary over what to do with it all and often resort to dumping practices not unlike those of fly-by-night haulers. Though promising to neutralize or recycle wastes, pressured by greed they merely dump the chemicals into the ground, bequeathing them to posterity in an active state. Waste yards become danger zones with fumes hovering above and fires erupting on the surface. Sewers are contaminated, creeks reduced to open cesspools. Disorganized and remarkably lenient, the government has looked on as at least a passive conspirator while these businesses, these poison merchants, have brought near apocalypse to numerous communities.

Corporate titles, liberally employing words such as "environment," "protection," or "ecology," attempt to convey a false sense of responsibility and social concern. So it was that the company in Louisiana where the boy was killed, and where waste pits were located below the Mississippi's high-water level, called itself the Environmental Purification Advancement Corporation. Like an imported animal species let loose on virgin soil, the waste brokers have proliferated throughout the nation, settling into a lagoon complex in one place, a landfill in another, and focusing their attention on highly industrialized states—California, Illinois, Indiana, Ohio, Texas. Several of them—SCA Services, Waste Management, Browning-Ferris Industries, Rollins Environmental Services—have reached the proportions of conglomerates, gobbling up independent landfill operations and spreading themselves on vast sites.

A few of the waste companies operate admirably, developing imaginative designs for recycling, and constructing, maintaining, and monitoring safe landfills. On their properties waste-barrel contents are neutralized and incinerated, or recovered if it is profitable, and the dregs and cinders are placed in specially designed "secure landfills" that consist of leachate-draining sumps, compacted clay, and dike systems to prevent washouts. Unfortunately, as we have seen, few of the companies operate in this way, despite what their slick brochures proclaim. The more general rule is landfills excavated carelessly over poor sub-soils, in operations that are often surreptitious and illegal.

A common practice for disposers, especially small ones, is to collect enough drums to fill several acres of a lot, charge industries for nonexistent waste treatment, and then simply abandon the leachbeds and deteriorating drums for the taxpayer to contend with. There are enough of these cases to fill a formidable volume. A company ironically called Pollution Abatement Services of Oswego, New York, left about 8,000 drums at one parcel and hundreds of others at several other disposal points; no one is quite sure what is in all of them. The firm was founded by several professionals in Oswego who, after deciding that they could not operate properly and profitably at the same time, declared the firm "hopelessly insolvent" and simply left the toxicants near the shores of Lake Ontario, where the nauseating fumes permeate the immediate vicinity.

In the city of Lowell, Massachusetts, in the early 1970s, Dr. John Miserlis, a chemical engineer and former college professor, began a 5.2-acre waste plant which he called Silresim Chemical Corporation, a name derived by spelling his name backwards. Miserlis said the site was intended to be a model for the world to imitate, but in actuality the firm was run much as its name was spelled. Instead of being recovered, the chemicals remained in drums that were piled ten feet high, nearly covering the entire property. Fumes pervaded the premises to the extent that protective gear was essential to enter upon it. The surrounding terrain was blanched and stained, pathetically populated by defoliated trees. There were homes within reach of sickening odors, and some of the barrels sat only a short distance from the River Meadow Brook, which flows into the Concord River, a tributary of a major drinking water source, the Merrimack River.

It was not until 1977, after years of operation, that Lowell city officials learned that Silresim existed. The state, though, had granted Silresim the necessary licensing to operate, and then had failed to close the plant until much of the damage was already done. The city became aware of Silresim accidentally, while searching for the source of fumes that were causing illness among workers at a diversion-chamber construction site for a wastewater treatment plant, and also investigating reports that industrial odors were rising from toilets in nearby households, apparently because of solvents flushed into the sewers. When it stumbled upon Silresim, the site by then had accumulated about 1 million gallons of waste in drums and another 250,000 gallons in large bulk tanks, including plating and etching waste, solvent residues, and aqueous mixtures of various other organic and inorganic chemicals, which would cost the state about $1 million to remove. In 1977, having exhausted his funds because of the firm's rapid expansion rate and unable to properly handle and reclaim the incoming material, Dr. Miserlis declared bankruptcy. Soon after, a judge ordered the facility abandoned, despite protests from the founder that if he was given a chance, he would be able to put the plant in proper shape. "I had a dream of responding to a need which exists throughout the country," he said. "But I'm afraid I was ahead of my time." That flattering self-image obscured the simple truth that he had entered a business he apparently had neither the funds nor the knowledge to operate, experimenting with a delicate if dirty business that had made life hazardous for the public. Though not in the sense Miserlis had intended, it was a model for all too many others.

Sources at the United States Environmental Protection Agency told me of another firm, Seymour Recycling of Indiana, that presented an identical problem. According to one EPA investigator, the firm's parent company, Seymour Manufacturing, was attempting to sell the waste company because of its deteriorating condition and because it did not wish to assume responsibility for what it had accumulated. At one time about 50,000 drums were stockpiled, along with two small incinerators of questionable capability and reportedly leaking tank cars, a hundred yards from a school. The rural folks nearby complained of mysterious cattle deaths and stunted corn growth, which they associated with fumes in their air. When health officials tested snow discolored by the incinerator's plume, they found traces of lead and chromium. Wastes on the property were also suspected of containing C-56, cyanide, and a variety of organic materials; responsible firms such as IBM and General Motors had disposed of their wastes through Seymour. Only after political pressure intervened did the state's attorney general prohibit further shipments to the site. By this time there had been a series of explosions at the plant, including a gush of smoke from an erupted barrel that formed a doughtnut-shaped cloud in the sky, resulting in dozens of phone calls from baffled residents. According to EPA correspondence, the former president of Seymour had "abruptly left the scene" that same year, and there was no cash bonding on the facility nor assets left for a cleanup, meaning that revenue would have to be generated by taking in more drums before the yard and warehouse could be cleaned. Eventually the "plant" was sold to Chem-Dyne Corporation, a large waste broker based in Hamilton, Ohio, which served as sort of an intermediary between about 100 chemical manufacturers, including Monsanto Chemical, Allied Chemical, and Procter and Gamble, and 100 waste disposers, among them Donald Distler. (The wastes were placed in everything from an abandoned Titan missile silo in Idaho to a 2,200-foot-deep injection well in Oklahoma, or even shipped to Arkansas and Canada.) A reduction in the number of drums has finally begun at the old Seymour site, but only months and many complaints later.

Industrial poisons frequently are mixed with nonrefinable oil waste at salvage houses where the operators have little knowledge of toxicology and no equipment to test for what they are receiving. To the average person, the sight of contaminated oil sludge, despite a slightly caustic smell, would not be threatening. The same appearance can fool the unqualified salvage company. Other times, when the dangers are known, waste disposers are blinded by the money to be made and simply disperse the substances as if they were innocuous, thereby avoiding the expense of sending them off to authorized disposal facilities. To increase their income further, disposers have also been known to use the sludges on dirt roads in places such as northern Alabama, collecting a fee from those who wish to rid their property of dust. Or as happened in Furley, Kansas, recently, farmers have unknowingly rubbed poisonous oil on cattle to keep away flies, and this has resulted in the destruction of their herds.

Russell Bliss of Fenton, Missouri, was one of the oil "scavengers" who made a living this way. During a period of several months in 1971, he drew from a waste storage tank at the Northeastern Pharmaceutical and Chemical Company in Verona a residue from the manufacture of trichlorophenol which the plant used as a starting material in the production of hexachlorophene, an antibacterial agent once employed widely in household soaps and hospital washes. Bliss reportedly had subcontracted for this job from another firm, the Independent Petrochemical Corporation of Saint Louis, which had been paid $4,625 to collect the material, a price Northeastern found lower than what it had previously paid a larger firm to dispose of the useless matter. Bliss removed a total of 18,000 gallons of trichlorophenol and other residues, enough to fill six tank trucks, and placed it in storage at his own facility, where it was apparently added to discarded motor oils and lubricants culled from service stations in eastern Missouri and southern Illinois.

For all those initially involved, the transactions were economically sound ones. In the bargain, however, there lurked a significant drawback. As Northeastern knew, and as they claimed to have warned the others, the hexachlorophene by-product contained an extremely dangerous level of toxic materials—again, it was dioxin, in the part-per-million range. Experiments conducted with animals fed a diet containing merely a millionth of the amount of dioxin in Missouri soil later sprayed with the waste spelled death for 5 out of 8 Rhesus monkeys within a year's time; the initial symptoms included scaly skin, hair loss, and bone marrow disorders. In other monkey experiments, there were indications that even smaller dosages could be lethal, and in rats, it was suggested that 5 parts of dioxin for every trillion of food could lead to cancer. It was therefore important that quantities so high be handled with the utmost care, and that their resting place be secured against any environmental contact. The respect in which the potency of this substance was held was demonstrated in an episode that occurred during 1975 and 1976, when the United States Air Force attempted to discard dioxin extracted from the herbicide Agent Orange in a landfill near West Covina, California, but was deterred in its plan when state authorities forced the government to remove it from the state. The dioxin was then taken to a site in Arlington, Oregon, where a firm that specialized in the handling of hazardous waste planned to place it thirty feet below ground. Not even that was acceptable to Senator Mark O. Hatfield, who demanded that the dioxin be removed from *his* state and placed under military guard, for eventual incineration at sea.

Russell Bliss's handling was nowhere near so elaborate. On May 26, 1971, he sprayed the dioxin-laden sludge on a horse-breeding farm in the vicinity of Saint Louis. Within thirty-six hours, sparrows and other birds roosting in the barn rafters dropped from their perches, and died. To the mystification of those working in the area, the next several weeks witnessed the strange deaths of cats and rodents, all exposed to the dust-controlled farmland. Hundreds more birds were found dead. Of 125 horses kept at the stables, 85 had been exercised on the sprayed ground. The first horse death was recorded on June 20, and despite removal of the soil and replacement with clean river sand, the animals continued to lose weight, turn bald, and develop skin lesions, inflamed hooves, colic, and conjunctivitis until 1974, by which time the death toll of horses at the one farm had reached 48. No one knew exactly what was causing the frightening effect, only that its path included two other horse arenas and a rural roadway also sprayed that summer. Another 15 or so horses died of the same symptoms, and 12 cats, 6 dogs, and 70 chickens also succumbed. To a lesser degree, humans living nearby reported illnesses, and a six-year-old girl was rushed to the hospital with respiratory troubles and a hemorrhaging bladder. (The girl survived, but her mother, Judy Piatt, was bitter enough at Bliss to begin rising early in the morning and following him through his daily routine; she claimed to have photographs of chemical drums he had buried near a waterway. She told me she herself had suffered a stroke. A regional EPA official said that citizens' groups had also followed Bliss, claiming up to 40 other incidents.)

Not until 1974 did researchers at the Center for Disease

Control in Atlanta identify dioxin as the culprit. By that time soil laden with the compound had been used as grade under roadways and as residential fill, and oil tainted with dioxin had been sprayed on roads in Illinois. What was left behind at Northeastern remained in a black tank, and no one knew how to handle it.

Another example comes from eastern Texas, where, in 1979, motor oil contaminated with nitrobenzene, a solvent which may cause cancer, was illegally disposed of upon six roads. The chemical had been mixed with the oil by Browning-Ferris Industries, the nation's largest disposal firm. (A chemist who worked at Browning-Ferris testified that he was forced to resign when he objected to orders from management to blend in the harmful substance.) The toxic mixture was then provided without charge, and without warning, to contractors and even town commissioners, to be used as a dust suppressant. Those who lived near the improperly disposed oil complained of nausea and livestock losses. Similar incidents occurred in South Dakota and Minnesota.

At the EPA's Office of Solid Waste, I learned that the treatment—or lack of treatment—of waste oil is a scandal that goes back to World War II. At that time, hundreds of companies were in the business of re-refining used oil. But the high-performance car engines that were developed needed oil with additives, and the old re-refining processes were no longer sufficient.

But that, according to EPA sources, was not the only reason we began to waste massive quantities of used oil. After all, re-refiners could have adjusted their technology. However, as a glut of crude-oil supplies developed, and since lubricant oil is a high-profit item, oil companies were not exactly delighted with re-refinement, which cut into their profits.

"The companies put barriers up," says Hugh Kaufman

of EPA's hazardous waste division. "It was very interesting. The Department of Defense had a group that wrote standards for all petro products for federal and state governments. Everyone keeps to those figures: it's the only standard for purchasing. Any oil that doesn't make it gets squeezed out of markets. What the companies did was get a sweetheart deal to exempt re-refined oil—it was thrown out for use. Even if re-refined oil meets the standards, it still cannot be used." Kaufman alleged that the official charged with forming those guidelines later went to work for the American Petroleum Institute. "In 1965," he adds, "the Federal Trade Commission, based on lobby efforts, put labels on re-refined oil. That label discourages those who want to re-refine." When Kaufman protested such wasteful practices, he was defeated. As a result, while countries such as Germany and France have good programs for reusing oil and recycling other wastes, Americans continue to use oil once and then simply toss it away. According to federal estimates, in 1977 only 100 million gallons —out of 1.4 billion gallons used—was re-refined in only 25 plants scattered about the country. The rest was burned, used as road oil, or simply dumped into sewers, onto farmland, or indiscriminately into ground pits. And because industry does not always segregate its wastes, much of the used oil was mixed with toxins.

Identifying such compounds is a complex process, and only the most skilled technicians know what to look for and how to find it. Laboratory researchers employ the gas chromatograph–mass spectrometer system of analysis for their trackings and measurements because of its speed and precision. The gas chromatograph records the reaction of chemicals to elements in the column of the machine. Some chemicals take longer than others to travel through the column's matter; this characteristic is known as the "retention time." The chemicals in their routes create configurations that allow the operator to determine roughly what kind of substances he is dealing with. For more accurate identification and measurement, the mass spectrometer, often a tall black box with dials for ionizers and filament adjustments and switching red lights, is brought into play. As chemicals vaporize in the mass spectrometer, they disassociate and are run through a series of electromagnetic fields in a way that allows the machine to determine their molecular weight and, computing the "fingerprints" of various fragment movements, to compare them with a data bank of 25,000 or so catalogued compounds. The process can be visualized by imagining that the molecules are steel balls flying by a magnet. The attraction of the magnetism causes their trajectory to curve. By sorting out this spectrum of molecular "trajectories" and thus their weight, a mass spectrometer can determine the "fingerprint" of the compound. Searching for dioxin requires special preparation, and it was not even looked for in the Missouri episode until much later.

The larger waste disposal firms have more capital than small salvage companies to spend on such techniques and do not usually resort to devices as primitive as road-spraying. Those that are licensed and choose to follow the regulations properly and use the "best available technology" develop their dumps on stable ground richly endowed with relatively impermeable clay, place additional clay at the bottom and compact it for additional protection, and then line the landfill with a plastic sheet reinforced with nylon and known as a "hypalon liner." They carefully place waste drums in this container, segregating them by type so there will be less chance of a disastrous reaction. Withdrawal pipes and pumps are installed to draw out collections of liquid from the bottom of the landfill. When the dump is full, the top is graded with clay and soil so that it slopes from the middle and minimizes rainfall penetration. Finally, grass is planted on top to reduce surface erosion. Some of these landfills contain as many as 350,000 barrels, and there may be several of them at a large facility, situated alongside lagoons where incoming liquids are evaporated, landbeds where heavy metals and salts are precipitated, and facultative ponds where treated wastewater is readied for discharge into an outfall. Compared with past practices, the new landfill designs are a gift to the environment, but the cost of placing waste there can be high—from $10 to $150 a drum depending on the toxicity, or if it is in aqueous form, from $.18 to a dollar or more a gallon.

Disposed in "secure" landfills by firms properly equipped to handle them, chemicals nonetheless remain a cause for concern. However well kept the landfill, various compounds are certain to remain whose toxicity will persist for decades after the landfill has been closed and the present operators are long gone. No matter how good the clay sides and bottom, it is impossible to guarantee that there will not be a breach in its integrity, or that the land will not shift in the future, allowing escape passages, or that the surface cover will bear up to wind and rain. Nor can anyone be sure that the chemicals, reacting over time as the drums deteriorate, will not eat through the plastic liner. Who will maintain leachate-collection systems after a site is abandoned is also undetermined. While in theory the concept of a secure landfill seems to solve basic problems, many of them are improperly constructed and their bases allowed to pool spilled chemicals and leachate even before the final cover is placed. In other cases, the drums are randomly dumped into these landfills and the liner is broken by the sharp metal rims. However well they are separated, the chemicals will eventually mix together. And however slowly, they are still bound to leak.

Added to the inherent impossibility of structuring a fail-safe landfill is the problem of chemical concentration—the sheer volumes that collect at waste disposal sites. The huge new landfills now under construction will contain so much in chemical mass that should a serious problem arise, it would be financially prohibitive to excavate the waste. Because they often contain only the most toxic substances, collected from a plethora of industries, the contents are usually far more lethal than what would be found at the average landfill accommodating a single factory. While centralizing the landfilling of barrels decreases the exposure of the public at large, it can pose extraordinary risks to the adjacent communities.

In addition to handling waste from local factories, firms that specialize in disposal frequently import chemicals from other counties and indeed other states. It is not unusual to learn of a chemical drum that has traveled a thousand miles to a landfill, nor of waste arriving in New York State from as far away as Puerto Rico. At such massive sites, an explosion or fire could well grow beyond technical management, with the community adjacent to it suddenly finding itself in a deluge of toxicants and much of its land, practically speaking, contaminated permanently.

Accidents at waste disposal sites are legion. Along public routes leading to their entrances, tankers slop toxic material through dripping spigots, and collisions cause larger spills. Chemicals were spilled during a train wreck, tainting the countryside and fouling the drinking water in Point Pleasant, West Virginia. In 1977, the town of Kernersville, North Carolina, had to resort to an auxiliary water system for its 5,600 residents when chemicals from a waste disposal firm spilled into a twenty-two-acre reservoir. Six men, one of them a Baltimore County firefighter, were hospitalized in 1975 after inhaling noxious fumes at a commercial landfill near Dundalk, Maryland. Hydrogen sulfide gas, a quick killer, had been released when a reaction occurred on the ground. A fire and explosion in Shakopee, Minnesota, sent drums flying hundreds of feet into the air.

For these reasons, there has been a loud outcry in most states every time a waste disposal company attempts to locate in a community. It is an understandable reaction, but it has created a logjam of wastes and a dire shortage of licensed "secure" landfills. Protests were raised in Oklahoma City, Michigan, where a firm wanted to turn wastes to inert slabs and bury them; in Jackson, Ohio, where some 7,000 petitioners fearful for their livelihoods fought efforts by Browning-Ferris Industries to install a waste injection well and scour a landfill on a former dairy farm; in Hempstead County, Arkansas, where Delta Specialty bought 188 acres of land for the first planned hazardous landfill in the state; in Warrenton, North Carolina, where hundreds of residents opposed the government's opening a dumpsite nearby to dispose of PCB-tainted road soil; in Kansas, in Georgia, in New York, Illinois, and Maryland, and in Missouri, where residents successfully pursued a temporary restraining order to stop use of the state's first hazardous landfill just over a week after it was first approved by the state. The task of fighting large firms is a difficult one, for many of the companies have the blessings of governmental regulators and the clout to wage lengthy warfare. One of the firms meeting stiff resistance, Browning-Ferris, based in Houston with branches in a number of cities, was reported in 1978 to have 125 locations, and the year before, a gross income of $315.4 million. Just as important, the firm had considerable political influence, as did the other major outfits.

SCA Services of Boston, a disposal firm that by 1976 operated forty-two landfills covering 5,850 acres, is an outstanding example of a waste conglomerate that has had its problems both with public opposition and with environmental or other violations. Its first major problem with a site was in Wilsonville, Illinois, where one of its divisions, Earthline Corporation, owned a 130-acre spread. Neighbors became so upset at the prospect of spending the rest of their lives near a potentially dangerous landfill that they waged a fierce campaign that included hanging American flags upside down, recording the license plates of trucks that entered the facility, and placing on a utility pole in front of the property an effigy labeled "Method of Yesteryear." The state attorney general entered the controversy, and in what is expected to be a landmark case, Earthline was ordered by a Macoupin County circuit judge, John Russell, to stop its operations and remove what it had already buried there. Whether SCA, in future court hearings, will be able to retract the judgment is not clear; but the significance of such a public movement remains: soon after the decision, Governor James R. Thompson told the Illinois Environmental Protection Agency to stop issuing Earthline permits, and he also ordered a forty-five-day moratorium on the issuance of supplemental permits for waste importation to out-of-state haulers.

Equally dramatic was a citizens' fight in Bordentown, New Jersey, against an SCA proposal for a hazardous fill that would have been situated within 1,500 feet of a school and above the Magothy-Raritan aquifer, the major source of potable water for the area. Forming a group called HOPE (Help Our Polluted Environment), the people hired their own attorney, recruited chemists and geologists for expert testimony, and imported residents from the Love Canal to speak at public hearings when SCA's application was up for review. Adults jammed the auditoriums while their children chanted in protest outside. So efficiently did the people build their case that, in an administrative order issued in March 1979, the state's environmental department, over the signature of Beatrice S. Tylutki, a solid-waste administrator, determined that

> the reliability of the applicant to properly operate a land-based hazardous waste facility has not been demonstrated . . . that this demonstration hazardous waste disposal facility presents unnecessary and unwarranted burden to the public health, safety, and welfare. . . . Similarly, the records of the Solid Waste Administration indicate that on October 28, 1978, chemical waste was disposed at the Parklands sanitary landfill facility in violation of the conditions of its registration. This incident raises questions as to the ability of SCA Services Inc. or its subsidiaries to operate this facility. . . . For all the foregoing reasons . . . the application by Earthline Company for a demostration hazardous waste disposal facility at the site of Parklands is DENIED.

Aside from environmental problems, there were also serious questions about the firm's corporate ethics. The Securities and Exchange Commission had accused Christopher Recklitis, a former president of SCA, of illegally diverting nearly $4 million of company money for his own private and business uses, in the process implicating two other former SCA officers, Berton Steir, once chairman and chief executive officer, and Nicholas Liakas, former Northeast regional controller, as well as another man, Anthony Bentrovato, or "Tony Bentro," who had residences in Utica, New York, and Miami. Bentro was charged with selling a Utica landfill site to SCA at a grossly inflated price. Bentro, in addition, was indicted in 1975 in connection with a kickback scheme involving the Upstate New York Teamsters Pension Fund, along with New Jersey's Anthony "Tony Pro" Provenzano, a well-known *capo* in organized crime. Bentro was convicted, but later the charges were dismissed. While SCA itself was never publicly linked with the Mafia, there were persistent rumors of its association with such elements, not surprising in light of the history of regular garbage disposal operations that had come under the hand of organized crime in places such as New Jersey.

SCA employees had been charged with other misdeeds. The SEC's complaint alleged that the company had bribed town officials in Ohio who were contemplating the renewal of an SCA trash-hauling contract: "SCA paid approximately $16,000 to two Miami, Ohio, township trustees in 1974 in order to obtain a renewal of a three-year contract for waste disposal valued at $425,000. Such payment was made by means of a cashier's check dated January 24, 1974, purchased by one of SCA's subsidiaries, in the amount of $16,000 payable to two SCA employees who endorsed the check in blank. Such check was subsequently endorsed by the two township trustees. . . ." In secretly taped conversations, two officials of an SCA subsidiary in Trenton were reportedly named as having paid $75,000 to obtain a lucrative Philadelphia dumping contract. The two employees, John Zuccarelli and Leo Hughes, denied the charges. Meanwhile, a former company officer also was accused of making an illegal $1,000 contribution to then Massachusetts Governor Michael S. Dukakis. In western New York, where SCA had a sprawling complex that accepted shipments from up and down the East Coast and from Canada, the state's Bureau of Criminal Investigations, under the direction of the Niagara County district attorney, probed allegations that a former SCA officer had bribed a state senator, but this investigation did not succeed in leading to formal charges.

That companies with such questionable records were in charge of handling materials that could adversely affect public health was most discouraging, and one more indication of how poorly regulated the disposal of wastes was. Agencies often depended upon a company's word for what was transpiring on the site; they could not themselves check each incoming barrel for its content,* nor could they monitor each landfill for signs of subsurface migration.

So wastes would continue to be handled in dubious and dangerous ways. In unimaginable quantities they would continue to be stored in landfills or trenches or open liquid pits. They would continue to course through the soil and into the groundwater, or volatize into the air. Slowly but inevitably, they would continue to work their way into the human body.

* There were other rumors that a disposal company or two had started to smuggle narcotics in drums marked "hazardous" for transport throughout the United States and Canada.

13

UNDER THE MOONLIGHT

Jane L. Pleasant's first thought was that a wayward hog of hers had trudged onto North Carolina's Route 210 in front of her small white-frame house in Angier, blocking the lane for those few travelers who, at 12:45 this August morning in 1978, were straggling back from a late poker game or from taverns where discussions on the barning of tobacco were accompanied by bourbon and beer. She rose swiftly from her bed in the front room and stepped onto the porch to see why the car horn that had jolted her awake was still blowing. Looking toward the road under a slightly overcast sky, Mrs. Pleasant saw a station wagon with two men in it pull up behind a dark tank truck they had waved to the side of the road.

The station wagon was doing the honking, but not at one of her pigs. The vehicle had been trailing the truck in an effort to tell the driver that his cargo was leaking out of a back spigot at a rapid rate. As Mrs. Pleasant watched, a man from the passenger side of the station wagon began walking toward the tank truck. Before he got there, the leaking tanker was abruptly pushed into low gear and without warning pulled quickly away, down the road and into the darkness.

"Ah, let 'em go," said the car driver, with a heavy rural accent. "You can't help people, this day and time."

"Goddammit," said the passenger. "Let 'em lose their

damn mess if they ain't got more sense 'n that." With that the station wagon moved swiftly off the road shoulder, heading back to where it came from.

As the sixty-seven-year-old Mrs. Pleasant headed back into her house, confused by the ruckus but relieved that none of her hogs was to blame, she noticed an unpleasant, strongly pervasive odor in the night air. Her head began to ache, as it would for the next four weeks, and her eyes burned so much that she had to put Murine drops in them before she was finally able to get back to sleep. The next morning, her throat was scratchy and she felt nauseated.

"My sister-in-law said the next day, 'Did you ever in your life smell anything like this?' " Mrs. Pleasant recalled. "I said, 'Well, maybe it's someone spraying tobacco, and if the scent alone don't kill the worms, nothing will.' Then my grandson came over. 'You know what that is on the Black Creek Bridge?' he said. 'It's the awfulest thing you ever smelled.' Oh, Lord, the best I can describe it was like rotten eggs, ammonia, and vinegar—in between all that, that's what it smelled like. It killed weeds, grass, whatever, died like nobody's business. I stayed sick and my doctor said there isn't a thing I could do about it."

In front of Mrs. Pleasant's farmhouse, in front of sweet-potato fields, cow pastures, and tobacco rows for miles of highway right-of-ways in the state, was a brownish-black streak resembling motor oil. It was thick where the vehicle that spilled it had slowed to go uphill, and it was thin along the road declines. Although they were never certain the truck Mrs. Pleasant had seen was the one that was at least partly responsible, two weeks later authorities told local newspaper reporters that approximately 31,000 gallons of waste oil had been spilled on the land from communities forty-five miles north of the South Carolina border, west toward the Roanoke River, and on upward to the Virginia state line. And that in the spillage, which included the Angier area, were high levels of PCBs.

Initially, it was thought that 70 miles of roadside were affected by the suspected carcinogen, but as state laboratory workers tested dirt samples day and night for the next several weeks, the figure rose to 100 miles, then to 150 and finally to 210 miles of rich dark soil contaminated with the highly toxic waste. Four thousand letters of caution were hand-delivered to the residents along the tainted stretches, warning that because the chemical had been found as far away as 70 feet from the highway, they should not eat or sell vegetables and beef grown within 100 yards of the most severe contamination. Full of fear, some farmers complained that their migrant workers had sickened near the roadside. Others burned some of their crops, and Mrs. Pleasant halted use of her drinking well. It was not proved that any damage to health had resulted from the wanton and willful leakage. Nevertheless, a large portion of rural North Carolina had been jeopardized by what was perhaps the most blatant incident of illegal toxic dumping in the files of the United States Environmental Protection Agency.

Responding to the bizarre action, the state's Bureau of Criminal Investigation sent seven men on the case to hunt down the culprits while Governor Jim Hunt posted a $2,500 reward for information leading to the arrest of whoever was responsible for what one county commissioner called "an assault [on] our people and our environment in a senseless and cruel manner." By September, police had arrested Robert J. Burns, owner of Transformer Sales Company of Alleghany, New York, and his two sons, Timothy and Randall, in connection with the dumping. According to reports carried by United Press International, Burns had also operated a warehouse in

Alleghany that because of its sloppy condition was thought by the state Department of Environmental Conservation to pose a possible hazard to a nearby water supply; it was also reported, said a UPI dispatch, "that low levels of PCBs had been found in the water system of the Warren County, Pa. town of Youngsville, where Burns has stored a large quantity of PCB-contaminated oil in a rented warehouse."

At first, the three Burnses proclaimed their innocence in the North Carolina matter, but later they turned state's witnesses against the operator of a Raleigh firm, Ward Transformer Company, who reportedly had given Burns $75,000 to haul PCBs away. The three men admitted their guilt and could spend time in jail on both state and federal charges. A member of the state attorney general's office told me that the Burns boys were apprehended after a truck they were driving got stuck off a road in Halifax County. Rigged to the truck, he said, was a 750-gallon container attached to a pipe that went through a hole in the side of the vehicle so that its valve could be controlled by the driver opening the door on his side and simply reaching back to open the line. This was apparently done regularly over the course of several weeks, he said, in sequences of ten to thirty miles at a time, with the flow stopping as the vehicle neared urban areas. (It was alleged but not proved that other vehicles were involved.) In an attempt to make light of the misdeed, Burns vowed to the media that he was "willing to go before television cameras or whatever and stand in a barrel" of PCB waste oil. Such an act, whether Burns knew it or not, could have had serious repercussions. PCBs are known to cause severe skin rashes. In Japan, where hundreds of people had sickened in 1968 after eating PCBs in their rice oil, a survey was done of 13 pregnant women who had been exposed to the compound, and of these, 9 bore children with impaired liver function and other defects, including darkly pigmented skin. More discouraging for the people of North Carolina—and for those who ate crops grown in the region—was the fact that PCBs collect readily in human fat and do not leave the human system for years.

The Burns clan served as a classic example of what are known as "moonlight," "scavenger," or "midnight" dumpers—small, unethical businessmen who haul hazardous waste from industries at prices that often undercut those of the larger and more legitimate disposal firms, and then, to eliminate overhead and greatly increase their profits, clandestinely discharge the material into isolated gullies, forests, creeks, farm ditches, city garbage landfills, sewers, and lakes, or, as in North Carolina, along the roadside. Many are linked with organized-crime networks, most are never caught, and those who are apprehended are rarely charged fines or given jail terms commensurate with the tax dollars it takes to clean up their illegal mess and the harm they pose to the general public.

There are two main reasons for the development of a black market in hazardous waste: a dire shortage of specialized, government-approved landfills and treatment centers, and the lack of a manifest system that keeps records of chemical quantities from the time they are produced until their ultimate reclamation or disposal. Although California was one of the heaviest manufacturers of dangerous trash, since it had official dumpsites and a mandatory record-keeping program there were few cases of midnight dumping in that state. Conversely, many Eastern and Southern states had neither good dumpsites nor manifest systems, and they suffered the most frequent moonlight onslaughts. Sites capable of properly handling sludge were often so distant and the disposal costs so exorbitant that haulers thought little of the risk of getting caught, knowing full well that even if they were arrested, their actions would hardly be treated as a serious crime. To them, a vacant lot overgrown with weeds was most appealing, the perfect spot to unload a truckload of drums and let the taxpayers foot the bill for proper disposal. At the other end of North Carolina, on Summit Street in Charlotte several years ago, Mrs. George Herron was taking down clothes from her line one evening at dusk when she saw a trailer truck stop at a nearby lot and men begin to unload dozens of chemical drums. "They said they were just leaving it for now," her husband told me. "Nobody paid it much mind. They said they would pick it up in the morning. The next morning, there were two hundred or three hundred drums there. They're still there. Nobody's moved them." And nobody knew what was in them.

Chemical barrels and bulk liquids were being abandoned everywhere. They were found among the soft pines of South Carolina, in vacant houses in Louisiana, beneath elevated skyways in New Jersey, on the medians of a major highway outside Detroit, on abandoned lots from the Bronx to the state of Washington. In Illinois, there were twenty-two approved chemical landfills and supplemental sites that received approximately 2 million tons a year, but the state produced seven times that quantity of waste, and no one could say where the balance had gone. As late as 1977, arsenic, cyanide, oil, ink, and solvents were pouring into the environment of Indiana, a state with virtually no modern toxic-waste disposal facilities. In Hawaii, according to *Chemical Weekly* magazine, Energy Recovery Systems was at one time the only fully licensed chemical-waste disposal firm. The company estimated that it received 20,000 gallons a month of waste oil, yet 100,000 gallons a months of such materials was generated on the island of Oahu alone; the balance simply disappeared.

"There are so many cases of midnight dumping," said John Shauver of the Michigan Department of Natural Resources, "that it's impossible to single out any one as the most startling example." As Shauver talked with me about the situation in the Midwest, his department was in the midst of handling a highly toxic spill in Howell that typified the problem of promiscuous waste disposal. In this case, the Livingston County Highway Department had been told by fire inspectors that it had a fire hazard because of the storage of herbicides with flash-points of less than 200 degrees Fahrenheit. The department contracted with a licensed hauler, Howell Sanitary Company, to remove the old chemicals to an approved toxic dump. Instead, as I noted before, the chemicals, which included herbicides associated with dioxin and Agent Orange, were dumped into a burn pit fifty feet from a child's swing set and not much farther than that from a house, a marshland, and a fish pond. Michigan was forced to pay the cleanup cost when the company claimed it had not the resources to do this itself. The only reason the DNR found out about the situation was an anonymous phone call.

And this is a common practice?

"Yes," said Shauver. "It occurs so often it isn't even newsworthy here any more. In Detroit, a guy pumped wastes into a basement in an abandoned brick building and just left it there. There was another place, a regular warehouse, that a man rented part of and put chemical drums in. There were antique cars stored there, and their paint began peeling off from the fumes.

"These guys are cutting corners everywhere they can. Central Landfill in Montcalm County—supposed to be a normal garbage dump. But two 10,000-gallon tanks were secretly buried there. C-56 waste from Hooker was hauled there in 1977 and pumped into the tanks. The disposal firm was fined $18,000 and was forced to clean up the mess."

In Michigan, the state government has allegedly violated its own rules. During 1979, seven employees of the state's highway department, including the director, were arrested when the DNR discovered a department truck pumping waste salt over the side of a bridge near Brighton and into the Huron River, according to James Truchan of the enforcement division. But the case was later dismissed.

Under the high braces of the Commodore Barry Bridge, near the Philadelphia airport, the rundown towns of Marcus Hook and Chester, and the refinery stacks that dominate their skylines, is Number 1 Flower Street, an address that, one recent winter, showed both how very inappropriate the street's name was and how dangerous the backdoor game of moonlight hauling can be. On the oil-sodden shoreline of the Delaware River was a small tire-shredding business owned by Melvin Wade. But the old rubber did not occupy the whole yard; there were in addition thousands of metal drums stacked one upon another in an asymmetrical pyramid. Wade reportedly told investigators that he had been convinced by a hauling firm, ABM Disposal Company, to enter the business of cleaning and recycling chemical drums; they would pay him $1.50 apiece. There were no records of what he was taking in, and the contents of the barrels were simply dumped onto the ground.

Wade's illegal disposal area was bordered by Interstate 95 and a natural-gas storage center. One cold day the wrong chemicals got mixed with each other or somehow were ignited, and the volatile reactions touched off a fire that raged for more than five hours, forming a billowing plume of thick black smoke over the Commodore Barry Bridge and sending more than forty firemen to the hospital. Had the wind been blowing northeasterly, the toxic cloud would have rained its particulates on the city of Philadelphia. As it was, fallout surely settled on the river's surface while other toxicants of unknown potency floated on the updrafts, eventually to settle upon smaller communities downwind.

It was not the first time that ABM had been linked to occurrences of this kind. Keith Welks, the state's environmental prosecutor, told me ABM had been convicted of other illegal discharges, including the dumping of pharmaceutical wastes into a well near the multi-billion-gallon reservoir of drinking water for Philadelphia, as well as the emptying of a load of cyanide onto a street in Chester early one morning in 1976, from where the contaminants proceeded into Ridley Creek, threatening drinking water supplies downstream. In that case, according to press reports, the firm was fined $2,500. Franklin P. Tyson, president of the now-defunct firm, had also been indicted on twenty-two counts of bribery and one count of conspiracy in connection with the alleged dumping of industrial wastes in city landfills. A city streets-department employee was similarly charged in the case. Along with Wade, ABM was being prosecuted for the rivershore caper, which came as no comfort to those environmental officials who were required to devise a remedial cleanup plan and raise the enormous funds necessary for it.

The city of Philadelphia and its general vicinity have seemed to be a haven for such acts. Another dumper, Gustav Propper of Cornwells Heights, was sentenced to 28.5 to 58.5 months in prison for dumping 500 gallons of explosive and flammable solvents. Along with a number of companies he controlled, he was also linked to ten other cases in the Bucks County and Philadelphia areas.

Farther north, abandoned coal mines have been used

illegally to inter Pennsylvania wastes. The substances have been pumped in great volume through old air pipes leading into the network of tunnels that reach as far as 2,000 feet below the surface and extend horizontally for miles. Near Pittson, a part of the Wyoming Valley Coal Basin, during 1979 several thousand gallons of chemicals began to seep out of the mines each day, traveling through a water tunnel constructed in 1911 to drain the mines and then empty into the Susquehanna River. As one result, drinking water in Danville, sixty miles downstream, has been tinctured with dichlorobenzene, a suspected carcinogen. Yet Pennsylvania's maximum fine for anyone caught in the act of such dumping would be $300. Once Kin-Buc closed, much of New Jersey's hazardous junk was sent to Pennsylvania in quantities at times exceeding a million pounds a week—and to a state that had legal dumpsites for less than 20 percent of its own wastes. In a startling number of cases, the trash never arrived at the proclaimed destinations, and few of the dumpers were caught.

Favored spots for illegitimate dumping in New Jersey were the marshlands in the northern part of the state, the Pine Barrens, and the streets of Newark. So bad was the situation that environmental officials went before Congress to plead for increased federal vigilance, claiming, among other consequences, that the state's high cancer rate was almost certainly related to this kind of dumping. Contraband liquids were unloaded into street pits in the middle of the night, sopped up the next morning in municipal refuse, loaded into sanitary trucks, and hauled to landfills ill-equipped to contain the materials. The water supply for 42,000 people was closed for two weeks in July 1977 after a truck driver dripped PCB waste oil near the Runyon Watershed, within ready visibility of a sign reading "Potable Water Shed, No Dumping." The chemical was traced in the water and the wells will have to be monitored for decades to come. In a vacant lot in Newark, an arson investigator searching through a fire watched his shoes disintegrate from illegally hauled waste materials scattered on the field. As it turned out, the fire had been sparked by the chemicals in the first place.

After the closing of the Kin-Buc landfill, there was also a marked increase of illegal burials in sections of New England. Waste from such firms as Scientific Environmental Control Company of New Jersey began appearing at widely dispersed unauthorized sites. One of them was in Coventry, Rhode Island, on a pig farm owned by Warren V. Picillo, Sr., a man with an interesting legal past. Picillo had served time in federal and state prisons on gambling convictions, and in 1967 was called along with reputed crime boss Raymond Patriarca into a United States Senate subcommittee investigating organizations that allegedly were posing as legitimate news companies to disseminate race results. He had reportedly also been arrested in a raid on his farm during which thirty-four new and used trailer-truck tires stolen from the Sea Land Corporation of Massachusetts were found.

That Picillo was shifting his commercial ventures to the waste disposal business became obvious on September 30, 1977, when a fire marshal, investigating an explosion at Picillo's pig farm that had created smoke visible from downtown Providence, discovered numerous barrels of hazardous waste in a thirty-yard-long open ditch. These were found to contain aluminum hydroxide, cyanide, xylene, toluene, carbon tetrachloride, and one mixture that could ignite at under 80 degrees Fahrenheit. (Some of the drums were later taken to a disposal site in Niagara Falls, New York, owned by Newco Chemical Waste Systems.) A local chemist hastily warned authorities that some of the waste material, when heated, could turn into phosgene, a gas that causes death through suffocation—the same gas that was employed during World War I by the Germans to sear the lungs of enemy forces. In addition, an environmental engineer estimated that the contaminants could move underground at a rate of a foot a day, reaching a major wetland connecting into the Moosup Watershed in less than two years.

Not long after the explosion, state police arrested two Elizabeth, New Jersey truck drivers while serving Picillo a fire marshal's order to remove the drums. The drivers had pulled up to the farm in trailer trucks and, in plain view, had begun dumping more matter into the pit until they were stopped. The trucks were found to be registered with Scientific Environmental Control. After posting $100 bail each, the drivers disappeared. Two days later, a superior court judge, at the behest of the attorney general, ordered a halt to any additional dumping, terming the farm "a chemical nightmare." Fifty barrels were taken from the site with no one at the "farm" able or willing to say where they had gone; many more remained behind. "The site was bulldozed over while we were in the chambers trying to get a restraining order," R. Daniel Prentiss, a special assistant attorney general for this case, told me in July 1979. "Picillo wasn't arrested. There was no statute at the time to cover it. There is a civil judgment against him, finding him liable for a remedy, but we're not yet sure how much it will cost, and it hasn't been executed yet."

Picillo's son was annoyed with the authorities, who, he said, through the publicity brought upon the case, had hurt a garbage-collection business that he operated. As investigators inspected a pit at the farm one day, he stood on the muddy ground, according to the *Providence Journal*, smoking a menthol cigarette and offering to drop a match into the trench to demonstrate that the liquid was not flammable. "Just wait until I am well off your property," retorted one of the marshals.

New Jersey's debris was also found about twenty miles north of Coventry, in the Smithfield area. It was hauled there by Chemical Control Corporation of Elizabeth, New Jersey, the firm that created a monumental hazard in its hometown when it allowed highly toxic and flammable waste to collect near a warehouse in deteriorating containers. At the time, the operator of Chemical Control was William Carracino, who for other incidents in New Jersey, all involving hazardous waste, faced a possible twelve years in prison and $53,000 in fines. Interviewed by the ABC television network, Carracino explained that "the economics behind it is if you don't have to treat it, there's no cost except for transport." In this way, a hauler could purchase a truck, charge up to $20 or $30 for disposal of each drum, fill the trailer with the containers, and simply abandon the vehicle at the side of a road, escaping with the profits. Carracino suggested another alternative: "You can go out and rent a piece of property—rent it, don't buy it. Rent ten acres, twenty acres, get a permit to handle drums. Bring it on the property and just store them. As soon as the heat gets too great, just go bankrupt, get out of it."

The waste in Rhode Island was unloaded near a regional landfill owned by a William Davis of Smithfield, and while both men contended that it was not dangerous stuff, residents nearby complained of nosebleeds and nausea, and a chemist at Brown University said he identified in the concoction three potentially carcinogenic substances. Because it lacked the proper legislation, the state was unable to take strong and immediate action, and it had to formulate emergency regulations to halt the incoming flows. Indeed, it was estimated that 2 million pounds a month were being dumped in the tiny state, which lacked any proper landfill for it.

With Picillo's trench closed by the court and the Smithfield landfill exposed, attention shifted to a woodland in Plainfield, Connecticut, at the eastern end of that state. In January 1978, a hunter happened upon a truck dumping containers in a suspicious fashion and called in the state, who found plenty more. Hundreds of drums containing acids and solvents had been deposited in gravel pits on property owned by C. Stanton Gallup, reportedly the wealthiest man in town. Gallup, a past president of the American Baptist Convention and a man who, according to the press clips, also had a Little League diamond named after him, was indeed a man of substantial means. He was reported to own 6,000 acres of land in Plainfield and 22,000 acres in Maine, and he operated a number of lumber yards and other businesses. In addition, he operated a community water supply located a short way from the dumping.

The material had been taken to the site by Chemical Waste Removal of Bridgeport, Connecticut. Along with the waste pits a makeshift toxic septic tank was discovered, consisting of a buried dump truck filled with stones and loaded with urea-based waste. Exposed, Gallup pleaded "no contest" to charges of dumping toxicants without a permit, was charged a $25,000 fine, and agreed to pay for the remedial action, which cost more than $600,000. Among others arrested with him was nineteen-year-old Emmanuel Musillo of Staten Island, who was listed as Chemical Waste Removal's president and whose father, Charles, was doing four years at Danbury, Connecticut, for conspiracy in bank fraud. Another Musillo son, Michael, also involved in the waste business, had once financed with his father what was described in organized-crime literature as a loan-sharking operation. Charles Musillo had been seen in the company of the famed mobster Matthew "Matty the Horse" Ianniello. Emmanuel, who was given a six-month sentence for the Plainfield dumping, had operated, under the firm, a boarded-up warehouse near the Bridgeport Harbor that had accumulated such a precarious volume of unknown chemicals that the state obtained a court order to occupy the building and, with police on guard, began the tedious and expensive task of identifying the contents and shipping some of it for permanent disposal at Chem-Trol Pollution Services, another landfill in Niagara County. Others arrested for either trucking in or arranging for the discharge of wastes in Plainfield included John M. Granfield of West Haven and Dominic "Slats" Marangelli of North Haven. At one court hearing, a defendant in the case claimed that he thought the waste he was hauling was simply spaghetti sauce by-products.

In the Connecticut case, the suspicions of underworld involvement were enough to bring in the Statewide Organized Crime Investigative Task Force for assistance, and as in many other dumping atrocities across the nation, there appeared to be possible links to established families in organized crime. It was a lucrative illegal business. While it cost from $20 to $155 to bury a drum of toxic material in an approved, secured landfill, the price of illegal dumping in Plainfield was only $1.50 a drum. Many investigators believe the deals with the generating corporations are struck by the high echelons of crime, then contracted to the small-time runners at an enormous profit. Said Prentiss of Rhode Island: "I guess the haulers are lightweights in a heavyweight game. The initial contracting is where the biggies come in." Corporate entities specializing in hazardous waste change their names so often that it is nearly impossible to track down the principals. Detectives in New Jersey and Pennsylvania have said that some haulers clearly have had previous connections with organized crime, including the Gambino family of New York, and both states are currently investigating a number of cases where such alliances have been indicated. In some cases, witnesses have been fire-bombed or otherwise harassed.

Whatever the influence of syndicate bosses, it was the independent haulers, the farmers or recyclers who operated smaller fly-by-night disposal "facilities," who were creating the most immediate disturbances. And nowhere was this to be seen more readily than in the Bluegrass State, Kentucky. It too had become a favorite dumping ground for the scavenger operators, and it too lacked landfills geared to handle hazardous remains. Consequently, material from industries in and around Louisville and from states such as Indiana and Ohio ended their cycle in the woods and farms of the outer rural regions of the state, reaching such a degree of saturation that Senator Wendell H. Ford charged that the dumps "could turn out to be the most costly—and potentially dangerous—man-made disaster in our state's history."

Illegally dumped compounds had leaked through soil in Marion County, dissolving a water main and making faucets run dry; they had accumulated alongside highways and creeks, uncovered by sudden floodwaters; they were strewn in drainage ditches and found floating in brooks. Near one dumpsite in Hardin County, where the operating company claimed to have no knowledge of liquid burials in its pit, lived Mrs. Chester Goodman, who seemed to think otherwise: "I saw them hauling at night. The one with the chemicals they weren't allowed to dump was a tanker. We just walked up there and the people told us we weren't allowed. They seemed upset. The loads came in, well, after I went to bed. I called the health department. They came, but it did no good. They [the dumpers] denied they did it. They told us it was nothing more than cleaning fluid. Well, this caught fire on two different occasions. The water here, you know, during really rainy times it tasted like cleaning fluid, but they tested and didn't tell us what was in it. They came back for more tests, though. If you put it in a glass container overnight, you could smell it. And I had good water, before."

One of Kentucky's foremost clandestine dumpers seemed to be Donald Distler, president of Kentucky Liquid Recycling. According to investigators, drums he had collected were found in an old brickyard at West Point and in a crowded warehouse in Portland; 16,000 barrels were eventually removed to an unknown destination because, in the official view, they were creating a fire hazard. During 1979, barrels were found afloat in Stump Gap Creek near a field owned by Distler's parents, Mr. and Mrs. William Distler. Among other residuals, they contained the notorious waste from C-56, or "hexa." In an emergency action, the EPA removed most of the drums to drier ground and prepared them for shipment elsewhere, but the agency ran out of money before a complete cleanup could be accomplished.

Donald Distler steadfastly maintained that he had nothing to do with his parents' problem, preferring to blame the predicament on a friend and former business associate, Kenneth Shelton, who was hauling for a different firm. William Distler went to court to sue Shelton, saying that the first he had heard of the barrels was from the Federal Bureau of Investigation. Shelton had a different version of the events. As related in testimony printed in the *Louis-*

ville Times, Shelton said Donald Distler came by one night and said, "Let's go for a ride. I want to show you something."

> He took me by the property, which he said was his property. It was real late in the evening. The sky was full of smoke. . . .
>
> He said, "Look over there. What's that?" I said, "Well, it looks like the whole country's on fire." And he said, "No. I'm burning waste." . . .
>
> I told him he was nuts. . . . He said he was burning liquid waste. And I asked him how, and he said he dug some trenches and he was burning it there. . . .
>
> Wasn't too much after that, I just went home.

Whoever it was that put the toxicants there created rather unpleasant working conditions for those laborers charged with pulling them out. Several young men wrestling the drums from the snow-filled creek claimed the job gave them headaches and rapid heartbeat and probably caused one worker to cough up blood.

That was little in comparison to what befell workers in Louisville's wastewater treatment plant nearly two years before, after Donald Distler poured hexa waste down a city sewer, where it collected in grit cleaners and other equipment in its path. According to an EPA report, thirty-four plant workers, engulfed by the resultant fumes, suffered acute physical effects of varying severity, including pulmonary edema, respiratory distress, nausea, memory loss, eye and skin irritations, and abnormal kidney and liver function. More than four miles of sewer were so seriously contaminated that the treatment plant had to be closed for two months, allowing 100 million gallons a day of untreated sewage to flow into the Ohio River. The residues were gauged at the extraordinary level of 100 parts per million in the sewer water. There is not even a federal standard set on safe consumption levels of C-56, but it is generally acknowledged that any amount above a single part per billion is dangerous to the human body. Thus, when officials in Mount Vernon, Indiana, downstream from the plant, learned that their drinking supply had been infiltrated at levels reportedly above this "standard," they temporarily shut down the community's water source. In Evansville and other towns, drinking water was treated with activated carbon to filter out the contaminants.

The first incident occurred on March 26, 1978, when plant workers attempted to clean a sticky yellow substance from their equipment and, in so doing, unleashed a noxious blue haze that quickly engulfed them. A year after the incident, one worker continued to suffer memory lapses; the EPA incident report said he would sometimes feed his cows twice in an evening, forgetting he had done so shortly before. Robert W. Keats, the attorney for the sewer district, told me that fortunately most of the effects had disappeared in the men. "As far as future effects—well, this is a carcinogenic substance," noted Keats. The attorney said the material apparently had come from a plant owned by the Velsicol Chemical Corporation and was brokered to Distler by a firm called the Chem-Dyne Corporation, which was also mentioned in courtroom testimony concerning the Distler farm and Stump Gap Creek. Distler was eventually convicted, but that did little to extract from the bodies of the plant workers a pesticidal residue so persistent and potent that officials had for a while considered hauling the contaminated sewer sludge out to sea on a barge and burning it at high temperatures. Nor did it repay the taxpayers the $1.25 million needed for an immediate cleanup of the lines. The charges were misdemeanors, carrying a total maximum sentence of two years.

To the east of the William Distler farmland, there was another disposal area whose dimensions approached that of a disaster. This time Distler had nothing to do with it. It was a 25-acre piece of property near Brooks, Kentucky, about five miles from Shepherdsville, owned by Mrs. Arthur L. Taylor, whose late husband had operated what was euphemistically known as a drum-cleaning business but could more accurately have been described as the most outrageous illegal dumping land in the country. Located in Bullit County on five acres of the Taylor land, the dump became known in early 1979 as the "Valley of the Drums."

The Taylor property presented environmental regulators with a grim collage of statistics: strewn topsy-turvy on the woodland were between 15,000 and 100,000 drums containing 197 chemicals. When a team of reporters from the *Louisville Courier-Journal* investigated, it found that among the wastes were xylene, used for solvents and dyes and dangerous as a fire hazard; isophorone, a solvent that is highly toxic when ingested and damages the cornea of the eye; naphthalene, which is used in moth repellants; and perhaps most dangerous of all, pentachlorophenol, a wood preservative usually contaminated with dioxins. In all, at least ten substances were associated with cancer in laboratory rodents, and there were also phenols, which in concentrated form could have caused death to humans through skin exposure. Mostly, though, the wastes were from the manufacture of paints. The drums had been accepted between 1976 and 1978 at the price of 75 cents apiece, then casually pushed into pits, emptied onto the soil, or, as one former Taylor employee told a reporter, dumped into nearby Wilson Creek. The most popular measure, it seems, was to simply heap the containers across the hilly terrain, allowing the chemicals—red and purple, thick like molasses—to ooze into murky puddles, turning trees into blackened logs and killing other forms of vegetation.

Arthur Taylor was known as a brash, even arrogant man who fought both with lawsuits and with more violent methods. In person, as the horsetrader he was, he spoke candidly. At one juncture, the "Valley of the Drums" had been a horse arena, but that venture held no lure for big profits. He had been a junk dealer, too, before he got into "hauling—whatever people wants hauled." One of the neighbors who did not fare well by Taylor's "hauling" was Joseph Vanhoozer, who told me he moved away from the land because of heated disputes with Taylor. Vanhoozer said, "He was getting a lot of paint and dumped it in a pond above my property. The hogs got in it. They came back solid red. That ditch down there—he dumped so much in there, my land, there were times the ditch, the water supply for my hogs, it was just a pool of paint. There was no water there. When I found out what the situation was, I took them up to the hill and fenced them in. I had some that were ready to go to market, and it would have done no good for them to drink paint, to get their water from there. There were barrels down the creek. They did it all the time. At night, especially at night, there were explosions back there. It sounded like a fort. There were times when it smelled like ether or rubbin' alcohol or whatever. It was awful uncomfortable at my place when they were dumping. And nobody helped. You better believe that. The environmental people got into it, but frankly, I was a bit rude. I don't care now. I moved away because of it. I had to take him to court, only lawsuit of my life, because of him imposing on me."

Government agencies not only neglected to respond to

residential concerns but also soft-pedaled recommendations of their own investigators, some of whom were apparently appalled by what they saw. An inquiry by the *Louisville Courier-Journal* showed that the state had first inspected improper dumping on Taylor's land in 1967 (seemingly before the heavy discharges took place) and then returned in 1976, 1977, and 1978, each time seeing what must have been quite obvious problems. In 1976, the state filed an administrative complaint against Taylor, and according to the newspaper, a hearing officer judged him to be violating water quality regulations. But that officer did not file his report for two years. He "forgot" to issue it. The site remained open despite the urging of action by an inspector, who sent memoranda warning of "deplorable conditions" that were posing a "serious health hazard." And no legal action was brought about until after Taylor was in his grave. The delinquency in responding to what was surely the largest illegal drum depot in the East was further explained away, by another official, as an "oversight." Not much more could be said.

On March 2, 1979, with heavy rains washing against the rotting barrels, the EPA declared an environmental emergency and rushed to the scene a response team that constructed earthen dams, diversion trenches, and a pipeline system for leachate collection to halt a massive flow from inundating a small stream nearby. The watercourse, Wilson Creek, ran to a tributary of the Ohio River, the source of drinking water for those thousands of people residing downstream. PCBs had already found their way into the stream, but the emergency action was declared a success in averting more serious contamination. The government was not so successful, however, in devising a permanent remedy. No community wanted the waste brought into one of their landfills, and it looked as if the Taylor property, as well as other makeshift Kentucky dumpsites harboring barrels with names on them such as Union Carbide, Ford Motor Company, Monsanto, E. I. duPont de Nemours and Company, Ashland Chemical Company, and Chevron Oil Company, was about to become the final industrial graveyard.*

Neither was a complete solution to be found for the roadside PCBs in North Carolina. There was great opposition to government plans to move those to a landfill, too, and so, in front of Jane Pleasant's house, the streak of contamination remains, covered over with sprinklings of activated carbon and emulsified asphalt. She still worries about it and harbors some thoughts about moonlight dumpers. "I'm a widow lady, and if I'd stayed sick, I would have been a mess," she said, recalling the strange incident. "Not only for myself, but for the schoolchildren . . . I think of them, chomping around in it. We older people lived part of our lives. If the children get cancer . . . I'll be right frank to you about those people who dumped: they knew what they were doing! If they don't care more about their fellow man than that, well, why, it's terrible, they should be punished. I have a lot more thoughts than that, but it wouldn't do to say."

* One plan was for a local man, William Fluhr, to set up an incinerator business on the property and take care of wastes that way. There were misgivings about the ability of the proposed incinerator to complete the job satisfactorily, however. There was concern that it could not reach the proper temperatures. The incinerator was sold to Fluhr by Donald E. Distler, according to EPA sources.

14

THE GUILT OF GARBAGE

One week after they moved in, Cruz Albert and Esperanza Negron were perplexed to see eight automobiles and trucks pull up to their home and the men who climbed out of them begin probing their property and that of their neighbors. The men explained that they were making routine checks. The Negrons probably thought that this was just another instance of the strangeness of life in suburbia. For many years, the family had lived in the less spacious surroundings of a Brooklyn apartment, carefully saving their incomes, and now, in 1976, they had made the move into a four-bedroom, cedar-shingled house large enough for them, their daughter, her husband, and the grandchildren. The location was Holtsville, Long Island, at the foot of a dead-end street and near what the builder had told them was going to be developed into a park.

Several days after the first appearance of the cars, an employee of the town of Brookhaven appeared at their door, explaining that his purpose was to measure the household air for gas. Again, there were no immediate signs of major trouble. But by September, the Negrons and a couple of neighboring families had to evacuate their homes because potentially explosive quantities of methane gas had been found collecting inside. The source was that "park" they had been told about. In reality, it was a covered-over municipal garbage landfill, filled with food scraps, paper packaging, and mowed grass, and from this decaying, oxygen-deprived refuse the tasteless, odorless, and invisible gas was emanating through the ground.

The incident demonstrated that even the regular, nontoxic town dump common to nearly all communities can, like a chemical landfill, cause an unhealthy environment. Created by anaerobic bacteria as they consume organic matter, methane is the main ingredient of natural gas and is known, in certain parts of the country, as "firedamp" or "swamp gas." Each ton of refuse is theoretically capable of producing 1,500 cubic feet or more of methane. In extremely high concentrations, methane can kill those who breathe it; if it constitutes 5 to 15 percent of the air, it can cause fires and violent explosions. Precisely such an accident occurred at the National Guard Armory in Winston-Salem, North Carolina, in 1969, killing three men and injuring twenty-two others, while in Springfield, Massachusetts, and Nashville, Tennessee, buildings exploded as a result of the fumes from neighboring landfills. There have been dozens of cases of methane entering households or business premises, including several on Long Island. One of the more serious cases took place in Northport, near Mike Nasti's Sand Company on what was known as Town Line Road. "This started three years ago, when all the trees started dying," Mr. Nasti said. "I first noticed it when a large spruce tree died. I thought it was because something had been spilled on it. It was near the fuel tank. But then I noticed many trees dying along the road and I called the town. It got in the building, and they had to put in an exhaust system and put vent pipes in the ground. At one point they said it was to the point of being explosive. What we did was have a guard stay here so we could leave the windows open at night. We had a lot of complaints of nausea, etcetera. The girls complained of a lot of headaches. I really believe it takes a lot out of you. I had to get out of the office or I'd go home with a headache. I didn't know what it was. Our building is only 120 feet from the landfill, and that was the first mistake."

Aside from methane, there are other threats from municipal garbage areas. While in small amounts they are innocuous, large aggregations of decaying vegetation, junked cars, electrical capacitors, and metal household items such as cans and pots, as well as other refuse, can send substances into the groundwater that contaminate drinking water with carbon dioxide and hydrogen sulfide, or with chrome, zinc, lead, iron, and other heavy metals. As it does with chemical wastes, moisture percolating through a municipal plot often carries deteriorating particles of garbage down to the aquifer.

Because there is much more of this brand of garbage than of more exotic forms, and because municipal dumps are frequently situated in sand pits or canyons, where liquid migration is swiftest, daily household garbage may be as destructive as many toxic industrial wastes. In 1974, three wells in Islip, Long Island, were found to be contaminated with high levels of magnesium, iron, and dissolved solids from garbage leaching out of the town's landfill. The water turned laundry red and rice black. Each homeowner not only lost his investment in his drinking well, but was forced to spend $2,295 to be connected to piped water.

The sheer quantity of garbage Americans produce is staggering. Municipal solid waste from residential, commercial, and institutional sources amounted to about 130 million metric tons in 1976 and should reach 180 million metric tons by 1985, with the average person contributing, by the 1976 figures, 4 pounds of refuse a day, or 1,300 pounds a year. This is enough to fill the New Orleans Superdome from floor to ceiling twice a day. The refuse is disposed at more than 18,600 sites, covering well beyond a half-million acres of land. In some cases, tons of sludge from wastewater treatment plants and tons of agricultural and mining waste are added to the heap.

Municipal garbage dumps have been a favorite point of discharge for toxic chemicals hauled by moonlight waste disposers or by legitimate industries that choose not to spend the money needed for secure landfills of their own. In many instances such actions have been illegal and clandestine, but just as often they have been carried out in the light of day, with the permission of the municipality and in a fashion that violates no statutes. Officials who have inventoried common landfills in industrialized regions have more often than not found crushed chemical drums among the domestic rubbish, or leachate containing chlorinated compounds or dangerous solvents.

New Bedford, Massachusetts, illustrates the dangers. For years, its municipal fill received waste from two companies, Cornell-Dubiler Electric Corporation and Aerovox Industries, without anyone investigating the consequences. Among the wastes, it was later reported, were 500,000 pounds of PCBs. In studies conducted between 1976 and 1978, it was found that the PCBs—compounds that can resist degradation for perhaps as long as one hundred years—had escaped into the ecosystem close to the base of Cape Cod. They were present in the air near the landfill, in the sediment of Apponagansett Swamp, and in the Paskamansett and Acushnet rivers, and were working their way up the food chain through benthic (river-bottom-dwelling) organisms and the eggs of the herring gull. This is not to say that all of the contamination came from this single landfill. Any municipal dump may receive PCBs by way of discarded electrical transformers, fluorescent light tubes, carbonless reproduction paper, marine-base paint, and radio sets. PCBs were first introduced in 1930, and by 1970, when their serious health effects were first generally noted, they were being sold at an annual rate of about 34,000 tons. By that time, they had been traced in virtually every American river and even at the Arctic Circle, in the flesh of nonmigratory bears. Their great tenacity for remaining in the ecosystem was cause for apprehension in the town of Dartmouth, which drew its drinking water near the southern end of the New Bedford dump.

In other places, investigators have been forced to stand vigil over trash heaps to keep the nocturnal haulers away, or even to dig into landfills and remove drums of dangerous chemicals, initiating all manner of new detective work. The quantities do not have to be huge to create a hazard, as was witnessed in Lebanon, Oregon, several years ago. A 6-inch by 4-foot cylinder of larvicide, manufactured in the 1940s, had been slipped into neighborhood trash, and when the sun had heated it sufficiently, it leaked an especially caustic form of tear gas, sending four firemen to the hospital with burning skin and eyes. City landfill operators have the disconcerting habit of spraying their access roads with contaminated waste oil to control dust or of allowing the burial of substances that are not inherently dangerous but present a toxic hazard in large enough amounts. In 1978 Iowa City gave the Procter and Gamble Company the right to landfill up to 10,000 pounds a day of bad toothpaste. It contained fluoride, which is perfectly safe for everyday use, but which in concentrated form can cause bone disorders and would therefore not be a healthy constituent of groundwater.

Economic and political expediency has led many communities to consider property for landfills that is inadequately sealed against rainfall and exposed underneath to flowing aquifers. When the city of Kokomo in north central Indiana became strapped for dumping room, it looked to a 103-acre parcel on the outskirts of town which, though superficially acceptable, in fact was atrociously constituted for a trash pile. The land, owned by Waste Management, at the time the nation's second largest solid-waste disposal firm and a substantial political force, was located on strata of unconsolidated clay, sand, and gravel deposited there by the glaciers of the Wisconsin Age. It was suspected of being linked with an aquifer known as the Tudor Drain, which branched into Wildcat Creek, connecting into the public water supply. "I believe that some of this water comes into these city wells by way of the [site]," said a local geologist, Dr. Roger F. Boneham. "Therefore any leachate which contaminates the sand seams which run through the [site] will ultimately end up in our drinking water, not to mention the water of the private wells of the residents who live to the south and southwest of the [site]."

Despite such expert advice, the mayor supported excavation of the landfill, and the local and state governments refused to give up on the idea. After years of heated discussion, there is still the possibility that a major landfill will one day be opened there; the issue is now in the courts. Were it not for a group of protesting residents led by Holly Schafer, surely the landfill, like so many across the country, would have received approval after perfunctory review. "As a housewife I was concerned when my husband would wake at two A.M. or three A.M. and worry what a poor-to-marginal landfill would do to our water supply and farm," Mrs. Schafer said at a hearing. "As a mother I'm concerned that a poor-to-marginal landfill could affect the lives of our children—perhaps more than it would harm us adults. Since my two-year-old can hardly do much to help solve the problem, I thought I had better get busy and take time to learn more about the issue."

For many communities like Kokomo, there are few alternatives to opening a new dump on marginal land, as the space shortage for municipal collection points grows evermore severe while the waste grows even faster. Americans are consuming 50 percent of the earth's industrial raw materials although their country has only 7 percent of the planet's population. In an average year, Americans discard 60 million tons of paper, 38 billion bottles and jars, 76 billion cans, and $5 billion worth of metals. Much of the waste is recoverable—perhaps as much as 24 percent or even more—but the country has been reclaiming only 7 percent of its garbage while spending, in urban areas alone, $6 billion annually to haul the potentially valuable materials away for landfilling. That cost represents the third largest chunk out of the local tax dollar. A report issued to the National League of Cities as far back as 1973 stated the situation bluntly:

> The disposal of wastes and the conservation of resources are two of the greatest problems to be understood and solved by this nation in the latter third of the century. With almost half of our cities running out of current disposal capacity in from one to five years, America's urban areas face an immediate disposal crisis. . . . The markets for recycled materials are severely limited due to federal policies which favor the use of virgin materials and discourage, even penalize, the use of recycled materials. There are no market incentives in force for recycled materials; virgin materials, on the other hand, receive the benefit of federally established depletion allowances and capital gains and credits.

The first tactic of an attack on waste problems should be the slowing down of disposal rates of plastics, metals, and glass and the recovery of potentially valuable materials by the use of separation devices. Several states have already passed laws mandating deposits on soft-drink and beer containers, and while industry has protested, the statutes have met with wide approval. Oregon, the first state to take significant steps in this direction, has effectively banned the sale of all aluminum and steel cans and throwaway bottles, and has placed a deposit on those containers that can be used by more than one bottler. Since glass and metals make up 20 percent of municipal garbage, the measure has had readily visible effects on the garbage flow. A second major measure should be increased construction of recovery plants that ferret from the garbage valuable minerals and fuel sources. The systems, employing conveyor belts, trommels, crushing machinery, screens, and shredders, can channel glass, metals, oil, and minerals to individual compartments for reprocessing. At an EPA project in Franklin, Ohio, municipal waste was pulped with water and processed into a slurry to make fiber that can be used for roofing felt and as fuel for electrical generation. Reusing garbage saves energy and creates less water and air pollution than does the processing of virgin materials.

Energy can also be withdrawn from garbage. Theoretically, says the EPA, the potential energy recovery from solid waste in urban areas is equivalent to 400,000 barrels of oil a day, or enough energy to satisfy the nation's commercial and residential lighting needs. Countries such as Denmark, Switzerland, Holland, and Sweden already recover more than 30 percent of the energy contained in their trash, while in 1977 the United States could manage only 1 percent. On numerous drawing boards are contraptions into which pork rinds, old newspapers, potato peels, and other household waste could be fed, and, after assorted mulching, produce combustible oils, gases, and charcoal. The technique is known as pyrolysis, and since 1970 more than fifty individuals and corporations have applied for patents on variations of the process, with several now in the stage of pilot operation. There have been notable failures among many of these attempts, but that pyrolysis will soon be a viable way of producing energy is no longer disputed. Energy can also be extracted in the form of methane gas from garbage that has rotted in a landfill. Wells are drilled into the landfill, and the gas is collected, refined, and piped as a supplement to other fuels. In California, where the nation's largest landfills are piled up in steep canyons and therefore present the best opportunity for methane gas collection, a project has commenced in Mountain View to tap gas from decaying refuse. It has met with enough initial success to draw a contract from the Pacific Gas and Electric Company.

With resources dwindling and common trash increasingly encroaching on populated areas, the time has come to recycle the garbage or find better ways of destroying it. The alternative is a continuation of the insane spiral which begins with throwing away usable items, depleting nonreplenishable minerals and metals, importing more oil, contaminating more groundwater, and chasing more families like the Negrons away from their cherished homes.

15

RADWASTE

An article that appeared in the *Louisville Courier-Journal* on October 26, 1978, began as follows: "The head of the Maxey Flats Radiation Protection Society has asked that radiation tests be performed on a woman who lived near the nuclear disposal site and who has cancer. Marjorie Denton, a member of the subcommittee on site status and improvement of the Special Advisory Committee on Nuclear Waste Disposal, said yesterday that John Hay asked her to discuss with the committee reports that several area residents had developed cancer in the last two years and that three of them died."

Maxey Flats is located on a ridge in the Knobs region of Kentucky, where a few scattered farm families live in the hollows. Where there are trees, they are usually oak or hickory, and underneath them are strata of green shale and limestone, with interbeds of gravel, sandstone, and hardened silt. Between 1963 and 1972, approximately 3.7 million cubic feet of "low-level" commercial nuclear waste —packaging paper, irradiated tools and glassware, and plastics—were buried at Maxey Flats in shallow trenches. Among the contaminating items was plutonium 239, the extremely poisonous, radioactive by-product of nuclear fission, which lasts 24,400 years before it is reduced to half of its strength. The wastes were shipped there by truck, in 55-gallon drums or in wooden and cardboard boxes, logged and surveyed at the entrance gate, then backed up to a system of trenches and randomly dumped twenty feet into the ground. Particularly large items and those substances emitting high levels of radiation were emplaced by crane, and at the end of each day, the ditches of hot waste were covered with dirt or tarpaulin.

The design of the Maxey Flats facility was as good as or better than that of many sites used for such purposes. The trench floors were sloped 1 degree to a sump pump and standpipe, supposedly to prevent leachate formation. Once filled, the ditches were capped with a minimum of three feet of soil, clay, and shale, mounded to discourage rain infiltration and planted with fescue grass and clover as a safeguard against surface erosion. At the time, these measures were considered more than sufficient for long-term containment of the dangerous materials. After all, as a United States Atomic Energy Commission report said, "chemical and physical characteristics of plutonium are such that migration in soil or groundwater is unlikely." Yet in 1972, when the Kentucky Department for Human Resources made a routine radiological survey of Maxey Flats, it detected elevated levels of radioactivity near the disposal site, which indicated precisely that form of migration. Upon further study, government technicians found that radioactive particles—tritium, strontium 89, cesium, and plutonium—had moved tens and hundreds of feet from the disposal trenches in less than ten years. "While EPA scientists are confident that at the present time this movement of plutonium and other radioactive materials does not present a public health hazard, the potential long-range impact of these contaminants is not known," declared the EPA, in a subsequent press release dated January 14, 1976. "Burial of radioactive materials in shallow landfills has been permitted since scientists believed that this material would not move to the surrounding environment during its hazardous lifetime. The problem identified in the [Maxey Flats] report reflects EPA's concern with low-level radioactive waste disposal practices involving shallow land burial facilities in humid areas in the United States." Although it had been said that it would not happen, radioactivity had apparently infiltrated surface runoff, penetrated the groundwater, or seeped through rock fissures and joints to surface outside the disposal grounds.

There is, as yet, no conclusion to the Maxey Flats problem, nor have the final chapters been written for similar radioactive-waste sagas in dozens of states where such garbage has been insufficiently landfilled. The lone certainty is that, while not comparable to the toxic chemical problem in terms of volume and acuity, past and current burials of refuse from weapons manufacture and nuclear reactors, as well as from other sources of smaller quantities, also present a substantial threat to the ground and surface waters and thus to the health of the citizenry. Like organic chemicals, "radwaste" will take future tolls if the hot dumpsites are not soon cleaned up. Because of the phenomenally high monetary costs of decommissioning a nuclear waste site, which can range to more than $1 billion for a large dump, it will be years and perhaps decades before this can be done, and though materials are and will be handled in increasingly improved fashion, fissionable wastes will continue to be placed underground, since, despite some fantastic proposals, there seems nowhere else for them to go for now. After 5,600 federal studies on the disposal problem, no final, guaranteed solution has been reached, and it would be folly to expect one soon.

Radioactive spoils piles have been growing at an exponential rate since the onset of the nuclear age thirty-five years ago. Each year, the average nuclear reactor produces 30 tons of unusable matter, with the larger plants generating 10,000 gallons of high-level liquids a year. Much of the waste comes from spent fuel rods, the pencil-thin metal containers holding the uranium pellets that power reactors. Once their energy has been used, the rods are cooled in water basins and then cut into small pieces, exposing the pellets for treatment with acid, a process by which some uranium and plutonium can be recovered. Such a chemical extraction and cooling process results in radioactive wastewater. By late 1978, there were about seventy reactors operating in the country, having generated in excess of 3,400 tons of spent fuel materials up to that point. Taking into account the 9 million cubic feet of high-level wastes from weapons production since World War II, this means there was theoretically enough radioactivity in storage to seriously contaminate every gallon of water on the face of the earth, and enough to administer a lethal dose to each man, woman, and child. The General Accounting Office has predicted that at the end of the century, we will have to contend with 265 million gallons of high-level waste and more than a billion gallons of the lower-level variety, much of which can be only temporarily stored today.

There were other commercial nuclear waste receptacles at West Valley, New York, Barnwell, South Carolina, Sheffield, Illinois (in the state at one time handling the largest volumes), Beatty, Nevada, and Richland, Washington—sites that, like Maxey Flats, involved the use of shallow trenches. By-products from the weaponry programs or various reactor fuels were sometimes left in deteriorating metal barrels barely covered with topsoil and unmonitored for ground contamination. Nor was Maxey Flats the only site that had produced fears of high cancer incidence. In 1978, the *Rocky Mountain News* conducted a door-to-door house survey in Broomfield, Colorado, and found that, in the 47 residences, cancer had struck 13 people during a twenty-year period in a two-block neighborhood where people had been complaining about a high leukemia rate. There was no proof that radioactivity was the cause, but it was a fact that the neighborhood drew its water from the Great Western Reservoir, whose source was a creek that meandered beside the Rocky Flats nuclear weapons plant two miles upstream. In 1973, the state health department took samples from the reservoir and discovered, to its dismay, ten times the normal background level of tritium, the suspected result of a spill at Rocky Flats. Two years after, another department report related forty times the normal level of plutonium in the reservoir's bottom sediments. While the state and federal governments saw no health threat in the accumulations, Dr. Carl Johnson, then county health director, assailed that viewpoint. He argued that plutonium, a more injurious substance than radium, was indeed a hazard at the current levels because, when it combined with chlorinated tapwater, its ability to concentrate in human flesh was greatly enhanced. And he asserted that in contrast to the state figures, a 1973 survey had shown two thousand times the normal level in Broomfield water, clearly cause for official concern. Even before his charges, government regulators had discovered that since 1958 the company operating Rocky Flats had stored oil contaminated with plutonium in 55-gallon drums on pallets near the central facility. The drums had ruptured and corroded, leaking their contents into the soil of the surrounding terrain. The contaminated parcel had to be covered with asphalt and the drummed waste sent to Idaho for safer disposal. At the same time concern was being expressed over Rocky

Flats, residents and town officials in Model City, New York, were growing apprehensive about what they perceived as unusual incidences of multiple sclerosis and breast and liver cancer near a twenty-two-acre radwaste depository called the Lake Ontario Ordnance Works. Located there, under conditions of poor security and questionable maintenance, were 18 tons of hot sludge, a one-acre mound of ore tailings, and a 165-foot-high silo—once a water tower—that contained radium-rich blends left over from the Manhattan Project. Drums had been haphazardly buried in trenches so shallow that wind and rain had exposed some of them. Levels of radioactivity were detectable in the adjacent drainage ditches running north to Lake Ontario, and in the air in the form of radon gas, which was released from a top vent in the tower.*

Fifty miles to the southeast of the Lake Ontario Ordnance Works, not far from the half-million people who resided in Buffalo, was the West Valley complex, a sprawling stretch of lonely building shells separated by thick brush and trees across a 3,331-acre tract. In 1966, National Fuel Services, a subsidiary of Getty Oil, began operating the facility as the nation's first—and at that time only—com-

* Upon inquiring into this case, I located a man, James Schmidt, who reported that his father, a safety and maintenance director at Lake Ontario Ordnance Works from 1965 until shortly before his death in 1972, had died of cancer of the intestines, spleen, and liver not long after climbing into the strange silo for an inspection. His dosimeter badge, according to later reports in the *New York Times*, was said to have gauged a lethal dosage, but the newspaper could not track down the badge, and the records of Schmidt's father's exposure had been confiscated by two men who approached his mother, saying they were federal agents. "They showed her badges and one of them pulled his jacket back to show her his shoulder holster," Schmidt said. "She was so scared she didn't ask to see his identification cards, and she doesn't know what agency they were from."

mercial-waste reprocessing center. Unable to operate at a profit or control the spread of radioactivity, NFS closed in 1972, a technological failure. But by this time it had already reprocessed 625 metric tons of spent nuclear fuel, leaving behind 600,000 gallons of high-level liquids that, although they will remain toxic for the next 100,000 years, are stored in a tank equipped to last for only the next several decades. While the state and federal government are wrangling over who should pay the $200 million to $1.2 billion cleanup costs, the poisonous residue periodically leaks into Cattaraugus Creek, a tributary of Lake Erie, the source of drinking water for Buffalo and dozens of surrounding communities.

The most notorious ground leakage of radwaste occurred at Hanford, Washington, where some of the nation's mightiest firms—Westinghouse, Exxon Nuclear, and Rockwell International among others—have operated, and where the plutonium was produced for the bomb dropped in August 1945 on Nagasaki. For years, Hanford was the largest nuclear dump in the world, topping the list with more than 50 million gallons of radioactive liquids, kept in corroding vessels and tanks. In twenty years, close to half a million gallons of radwaste had migrated outside the man-made barriers, and another 31 million gallons of low-level weapons debris, containing at least 190 kilograms of plutonium, had been dumped directly into the soil, prompting fear at one juncture that the result might be an uncontrollable chain reaction. More frightening, there was one fifty-day period when, unbeknownst to site personnel, 115,000 gallons of high-level material leaked from a storage area. It was declared that no serious health threat to people was posed, but nonetheless the Federal Water Pollution Control Agency described the state's Columbia River, one of the nation's largest waterways, as "the most radioactive river in the world." The first week of October 1979, Washington Governor Dixy Lee Ray shut down a commercial nuclear dump at Hanford after the discovery of defective packaging and shipping methods. About the same time, two other states with nuclear dumpsites, Nevada and South Carolina, attempted to reduce or close down operations, leading to a dramatic shortage of dumping space for, among other sources, hospital nuclear units.

Many of the smaller radwaste burial grounds, like those of the chemical industry, are unmarked and therefore unknown until an incident occurs to alert new occupants of the site. So it was that, in mid-March 1978, workers for the L. B. Foster Company's steel-pipe fitting plant in Washington, West Virginia, a new $2 million facility along the Ohio River valley, created a violent combustion simply by digging into the ground to pour concrete fittings for a heavy machine. Sparks and debris were scattered more than fifty feet into the air, scalding the roof of the structure, and the ground popped and flared like Fourth of July fireworks. To the surprise of the owners, investigators were to determine that radioactive thorium had been buried at the site and, mixing with some highly explosive zirconium also landfilled there, had created a volcanic blend. The property had once served a facility operated alternately by the Carborundum Company and Amax. Both firms denied burying radioactive matter, an action that would have contravened the law, but an official for the Nuclear Regulatory Commission labeled the dilemma a result of "discreet burials." Inside, one portion of the plant had to be roped off when radiation levels were found to exceed federal standards, and Foster announced plans to relocate its buildings to a more hospitable environment.

About 35 burial grounds similar to the Foster site are known to exist in other parts of the country, and certainly there are many more that do not appear in any record books. Waste from large factories or small hospital laboratories have been strewn about landfills without regard to future effects. Radiation readings have cropped up above roads where contaminated slag has been used as fill and under parking lots near community schools and playgrounds. (In Niagara Falls, the abandoned 99th Street School was found to be surrounded by low-level radiation, as were four other city schools and the surface of the Love Canal.) In many cases, contaminated dirt has been piled at the town dump and forgotten, only to be hauled off, later on, by contractors looking for a good bargain in roadbed slag. Few warning signs have been posted on these heaps. In Attleboro, Massachusetts, in 1978, federal and state investigators found radioactive debris on Finberg Field, a public playground, when they were investigating another hotspot near an old public dumpsite. In most of these cases, the radioactivity was so minimal that it would take days of physically sitting on the radioactive spot to receive a dose comparable to that of a hospital chest X-ray, but in other cases the levels, and the potential long-term effects, were not determinable.

For years, it was thought that low-level radiation from sandlike uranium mill tailings was not dangerous to human health. Uranium is commonly found in small quantities in sedimentary rocks. The ore is crushed, ground, and dissolved, the uranium extracted chemically, and the resulting tailings piled near the mill or elsewhere. By the late 1970s, when it was realized that perhaps even minute radiation might be harmful over an extended period, more than 140 tons of uranium ore bits had accumulated in the United States, or enough to stack 11,000 feet high over a single football field. The tailings contained small amounts of radium and thorium, materials discarded during the extraction process for nuclear fuels. In several of the Western states, where there had been 22 mills in the past and where at least 16 were currently operating, the waste material was used, not just for roads and parking lots, but as a building material for homes, schools, churches, businesses, and other structures. What this meant was that potentially carcinogenic sources of radioactivity were embedded in the homes and workplaces of thousands of Americans.

The problem was acute in Utah and Colorado. Within fifteen miles of Salt Lake City, a weed-choked railroad track was flanked by a 3.4-billion-pound collection of uranium residues, what Governor Scott Matheson described as "the largest microwave oven in the West." These tailings, unshielded from wind and rain and surrounded by a deteriorating fence, were reportedly accessible to the casual passerby. In Mesa County, Colorado, there was enough of a worry over what could happen to those living in the proximity of radioactive tailings to provoke a state health study in 1978 of persons residing in the county. After six months of study, the state concluded that there was no discernible cancer link—at least not yet—but at the same time its epidemiologists could not explain why the county's leukemia cases were twice the state average. Whether or not the cancer had been caused by the radioactive wastes, there was enough concern in Washington for deliberations to begin on multi-million-dollar projects to remove the tailings from the environments of Colorado, Utah, and other states. In New Mexico a dam owned by the United Nuclear Corporation and made of tailings collapsed on July 16, 1979, sending nearly 100 million gallons of low-level liquids into the Rio Puerca, which is relied upon by the Navajos and their farm stock as a source of drinking water.

Florida was another state affected. There was substantial phosphate mining there, and in the processes of extraction and fertilizer manufacture, masses of sand and gypsum tailings with significant radioactive levels were produced. Dams containing waste slimes from the fertilizer plants often leaked or collapsed, causing alpha irradiation of surface water supplies and of the sediments in the Peace River. An EPA memorandum issued August 15, 1978, stated:

> In the course of doing routine damage assessments, we have been alerted to the fact that a hazardous waste is being used commercially and may be endangering human health. Specifically, a material called phosphate slag, a waste product of a certain refining process in the Florida phosphate region, is being used as construction material. This slag has all of the radioactivity of phosphate ore, which is approximately 80 to 100 picocuries per gram. (The soon to be proposed EPA standard for radioactive wastes is five picocuries per gram.) Three Florida-based phosphate fertilizer plants generate a total of approximately 250,000 tons of slag per year and sell it to a single company which uses the material for paving roads, as material for railroad beds, as ballast, and for house roofing material. A possible 75,000 pounds a year used for roofing gravel is of particular concern, as residents of Florida's phosphate mining region are already exposed to natural radioactivity of two to twenty picocuries per liter in drinking water. Additional exposure to radiation from slag waste products such as those mentioned may be a serious hazard to human health.

The federal government predicts that it will be at least 1988 before a "permanent" solution to the nuclear disposal logjam is developed. If past experience is a barometer, it probably will be considerably later than that, if it happens at all. The ideas of what to do include everything from melting the wastes through the polar ice to shipping them toward the sun, but such measures are considered either too dangerous or economically unfeasible. More mundanely, scientists have been researching ways to solidify the wastes, package them in ceramic or glass containers, and send them into deep underground caverns, preferably dry climates where there are salt beds. They propose the solidification of radionuclides into crystalline or noncrystalline oxides that would be bound to other materials such as concrete or lead, and further containment by a ceramic outer layer or other sealants. The mineral monazite, because of its basic insolubility, is indicated as a material that might seal off radioactive wastes from the environment for millions of years. Additional safeguards might include final encasement in metal and then burial.

One idea for long-range disposal is a project called WIPP—Waste Isolation Pilot Plant—proposed for an area near Carlsbad, New Mexico. WIPP—which has raised great controversy among those who would have to live near it—would involve construction of tunnels and caverns dipping more than 2,600 feet below the earth's surface where, in vaultlike rooms, the canisters of wastes would be neatly placed, maintained, and monitored. This site was chosen because of the region's extensive salt strata, which, since they have been geologically stable for hundreds of centuries, proponents of the plan believe would minimize migration of radioactive substances. Those who do not subscribe to the salt method argue, on the other hand, that small amounts of water contained in the salt could combine with impurities like calcium and magnesium and form a corrosive liquid that would eat through the steel containment barriers, or that the waste material would heat the salt and cause it to expand upward, thereby pushing the radioactive quantities deeper into the earth into a totally uncontrolled setting. They also say there is no proven technology for recovering such waste from deep salt beds in the event that it is someday necessary to do so. Problems are also associated with glass entombments. There are scientists who warn that glass is too soluble in water and is prone to attack by pressure and heat. While these ideas have been bandied about, the nuclear industrialists have spoken of cutting down the volume of radioactive waste by treating it with laser beams in a fashion that would separate isotopes and create recoverable nuclear fuel, or of sealing solidified waste into ten-foot-high steel canisters covered by a layer of cement and letting them sit on the surface, monitored and cooled, until a depository can be decided upon.

Even the "permanent" solutions being proposed do not address the long-range questions of environmental safety for the simple reason that they do not include technology for rendering the hot stuff harmlessly inert. In the end, that will be the only solution. Further burial or collection of dangerous waste, however sophisticated, still leaves underground caverns laden with an unapproachable nuclear concoction that could eventually seep into the potable waters or the atmosphere.* Some of these wastes could remain toxic for millions of years, an awesome fact when one considers that only 25,000 years ago Chesapeake Bay was a valley through which the Susquehanna River snaked its way to the ocean, and that much of the Northeastern

* Yet despite unsolvable problems with our own wastes, President Jimmy Carter announced a plan in 1979 whereby, in an effort to prevent its use in weaponry by certain nations, the United States would *import* the radioactive debris produced from their reactors.

United States was covered by glaciers. There have been reports that in 1958, a nuclear waste explosion in Russia's Ural Mountains caused the contamination of 1,000 square miles of land. If that account is true, it shows further the need of permanently destroying, not simply burying out of sight, the weapons debris and spent nuclear fuel that continue to accumulate near production facilities and generators or to be scattered haphazardly throughout the environment, increasing the chances of cancer and ecological damage over a wide expanse of territory. For these reasons, states such as California, Maine, and Wisconsin have declared a moratorium on the construction of reactors until a remedy to the waste problem is found. Unless others follow suit, not too many years will pass before radwaste joins with toxic sludges in a massive assault on human life and health. In fact, it deserves to be the sole subject of another volume.

16

BEYOND COUNT

Love Canal, people now realized, was but the precursor of what was bound to develop in industrialized regions throughout America. During the past five years, reports of damages from discarded chemicals and radioactive wastes had been piling into the file cabinets of regulatory agencies with an alarming regularity, and government was obviously at a loss to know what it should do. Hundreds of hazardous-waste depositories were unraveling out of technological control, portending, for the woefully unprepared authorities, the most serious environmental threat of the 1980s.

Soon after the Love Canal evacuation, Thomas C. Jorling, assistant administrator of the United States Environmental Protection Agency, requested of the agency's ten regional field offices a list of *known* sites where hazardous materials were stored or buried. The request had come hurriedly, on the heels of a newly initiated congressional oversight subcommittee investigation of the matter. It had also been sparked in part by concerns expressed at the Office of Management and Budget over how much future landfill mishaps would tax the national treasury. The regions were instructed to submit their lists by October 23, 1978. A full month later, when the tabulations were publicly released, it was obvious that the prospect was grim. The regions had identified 32,254 sites. Of these, it was surmised that 838 had the capability of presenting "significant imminent hazards" to public health. "Improper disposal of hazardous waste constitutes an extremely serious environmental problem," EPA's administrator, Douglas M. Costle, asserted at last. "This preliminary survey indicates that thousands of potentially dangerous chemical dumpsites exist throughout this country. For decades, we have been disposing of these chemicals without adequate safeguards. We've paid very little attention to where these wastes have gone, in part because we weren't aware, and in some instances out of ignorance, and in some instances out of sheer carelessness." Former Congressman John E. Moss, then chairman of the Subcommittee on Oversight and Investigations, chose stronger verbiage. "Toxic chemical waste," he said, "may be the sleeping giant of this decade."

Between thirty and forty years ago, the United States embarked on a course in which industrial processes and the production of food additives, fuels, fabrics, construction materials, fertilizers, drugs, and pesticides became heavily dependent on complex chemicals. Synthetic fibers and plastics were developed, and petroleum was cracked into by-products. Between 1942 and 1962, plastics production alone grew from 500 million to 7.8 billion pounds, and the United States supplied nearly 40 percent of all the world's chemicals. Adding to the momentum was World War II, during which vast federal funds were allocated for corporate research laboratories, providing a capital and technical base from which, after the war ended, factories were able to greatly expand. In their infancy, the chemical trade and other industries were able to burn their residues into the air, allow them to drain into lakes and tributaries, dump them at sea, or simply flush them into municipal sewers. While the number of factories and the quantities of waste increased at a frantic pace, the technology for treating unwanted by-products remained as primitive as when the boom first began. There were no profits to be made from destroying the residues, only from hurriedly manufacturing the chemicals.

Finally, legislation was passed to prevent wanton despoliation of surface water and air. After the 1960s, many airbound particulates had to be scrubbed from smokestacks and large quantities of toxic sludge extracted from liquid waste. These and the precipitates from distillers and furnaces had to be disposed of by other means. As a result, hazardous wastes were pumped into drums and tank cars and hauled to unused corners of plant properties or to off-site trenches and garbage dumps, without the interference of governmental regulation. There they began to accumulate to the point where pockets of buried refuse dominated large sections of every industrial landscape.

Many forms of industrial garbage were generated, but it was the category of "hazardous waste" that soon constituted a new environmental dilemma. The term means any by-product that poses a substantial present or potential threat to plants, animals, and humans because it is harmful and nondegradable and may be biologically magnified (accumulating in the fat instead of being excreted). Examples of hazardous waste include filter cakes, organic tars, and other compounds from the manufacture of pharmaceuticals, waste solvents containing halogenated hydrocarbons, residuals or raw materials from paint manufacture, dust from air-pollution equipment, still-bottoms culled from pesticidal units, and lead-tank sediments from petroleum refineries. Hazardous wastes are either flammable, explosive, corrosive, or toxic.

Toxic chemicals produce injury upon contact with or accumulation in the body; they enter through the skin, nose, or mouth and produce injurious effects upon the body's organs by stimulating or suppressing the normal metabolic functions. As I have said, they may halt proper oxidation of a cell, thereby inhibiting the flow of energy, or combine chemically with tissues and provoke them to abnormal and uncontrollable growth. A chemical agent may attach itself so tightly to an enzyme—one of those complex protein materials that speed chemical reactions throughout the body—that there is a severe diminution in energy production from food and in the capacity to repair worn-out tissue. When enough cells have been harmed or set on a wayward course, there are serious repercussions; if the damaged organ is a vital one, such as the heart, liver, or brain, chronic debilitation will result, eventually followed by death.

Approximately 25 percent of industrial metals are toxic in high enough dosages. Metal wastes, emerging from processes involved in electroplating and in the production of dyes and pigments, fungicides and insecticides, and batteries, embed themselves in the bones or other tissues, thereby altering the normal function of the organs and in many instances attacking the nerve cells. In the environment they persist indefinitely. Organic wastes account for 60 percent of hazardous factory emissions. Among the synthetic organics are many of the plasticizers, solvents, and intermediates of manufacture. For their toxic punch, the most important organics are the pesticides and defoliants, which have been specifically created by man to kill forms of life and therefore would be more appropriately called "biocides." They are composed of the same elements, carbon and hydrogen, as human tissues, and because of this similarity to natural substances many of them are biologically active. Many of the chlorinated hydrocarbon pesticides, such as DDT, were added to the environment specifically because of their ability to combine with the outer fat of an insect and go on to destroy its nervous system. What we failed to anticipate was that these chemicals would react similarly upon other living things, including human beings.

As important as the specific actions of chemical agents are the sheer volume and variety of their wastes. During the period from 1943 to 1968, United States consumption of selected toxic metals increased by 43 percent. Production of synthetic organic chemicals has grown at a yearly rate of more than 10 percent since 1954 in the form of such materials as dyes, pigments, and the biocides. Meanwhile, each year approximately 1,000 new chemicals are added to the 70,000 already in existence, often more sophisticated than their predecessors and even more alien to the human system. It can be roughly estimated that during the past several years, upwards of 35 million tons, or 70 billion pounds, of hazardous waste have been generated per year, far in excess of the 10 million tons produced in 1970. And the trend is expected to maintain this same rate of increase at least until 1984, when more than 400 million tons of hazardous and semihazardous or innocuous industrial waste—more than twice the combined amount of municipal garbage and sewage—will be piling into the environment each year. To make matters worse, by that same year air-pollution equipment will be producing 83 percent more residues than it did ten years before, and heavy quantities of sludge from new wastewater treatment plants will be further overburdening landfill loads, in many instances with highly toxic metals and organics.

While the EPA did not know exactly where all the wastes had gone, it did have a general idea of how they had been disposed. Internal agency memoranda estimated that two-thirds of the residues had been discarded on plant property, which was shielded from public view but not from the environment. The rest went to more public settings. Prime among such disposal "techniques" was storage of wastes in unlined lagoons; 14 million tons went that course each year. What the EPA memoranda described as "environmentally inadequate land disposal" (in solid waste dumps or other unsecure surface impoundments) accounted for another 8.7 million tons per year of hazardous waste. Uncontrolled incineration, groundspreading, deep-well injection, sewer dumping, and the use of waste liquid for dust control on roads provided other means of unsafe discharge. Only a small portion—less than 7 percent—received proper disposal.

It was the time-honored habit of industries to dump their wastes onto land that had little or no commercial value: marshlands, abandoned sand and gravel pits, old strip mines, limestone sinkholes, or old construction staging areas. The great majority of dumps were not sited with hydrogeologic considerations in mind, and thus they tended to be in wet regions where leakage was most likely to occur. Unfortunately, it was in these wet areas, where precipitation exceeds evaporation, that industry was most heavily concentrated.

Approximately 70 percent of the nation's hazardous wastes are generated in the Middle Atlantic, Great Lakes, and Gulf Coast regions. A 1978 EPA draft report showed New York, Pennsylvania, Michigan, Ohio, Indiana, and New Jersey highest on the list of toxic landfills, with California, Texas, and Louisiana not far behind. Another EPA breakdown listed California as having the most industrial storage, treatment, and disposal sites (2,985). States such as Missouri, Tennessee, North Carolina, Kentucky, Colorado, Alabama, Oregon, Washington, and West Virginia belonged to the second highest category of producers. States with the lowest amounts were in the far

Midwest and Rocky Mountain regions, as well as in some sectors of upper New England. But virtually no state had a history devoid of landfill problems.

No dumpsite lacked the potential for harming the environment and those who lived in it. Groundwater contamination, surface runoff into waterways, air pollution via evaporation of chemicals and wind erosion, poisoning upon direct contact, fouling of the food chain, and fires and explosions were the most frequent disruptions. The examples were easy to find. In Delaware, a dumpsite accepting both municipal and hazardous wastes reportedly threatened a groundwater supply for 40,000 households. Nearby, in the grimy industrial town of Chester, Pennsylvania, 600 wells were closed because an unlined lagoon was leaking lithium, a metallic element apparently able to cause mothers to bear deformed young. In Belmont, California, 18 people were hospitalized, 1,500 evacuated, and 2 firemen permanently disabled with lung disease and brain damage when an old cylinder containing waste fumigants spewed its contents into the air. The deaths of 22 cattle in Franklin County, North Carolina, were traced to a trash heap containing calcium arsenate. In Batesville, Mississippi, 3 children were hospitalized for respiratory damage and coma as a result of exposure to the insecticide methyl parathion, which had been poured on the shoulder of a road.

Before Love Canal, the EPA already knew about 259 cases of well contamination, 170 surface poisonings of water, 17 air-pollution incidents, 14 fires and explosions, and 52 direct poisonings as a result of improper disposal. One EPA memorandum, poorly heeded, warned: "In the area of health effects, it is even more difficult to predict what, if any, future impact there will be, because most hazardous waste related injury to human health results from months and years of chronic exposure to trace concentrations of the toxicants. The health effects are insidious and sometimes impossible to trace back to causative agents."

Deep concern also developed about those new compounds created when several types of wastes were mixed together. It is well known that two chemicals, upon combination, can create a total effect greater than the sum of their separate effects, the phenomenon known as synergism. Hydrogen peroxide in amounts above 1.5 parts per million, combined with 1 part per million of ozone, was found to be lethal to some animals, whereas the peroxide by itself produced only slightly toxic effects at nearly two hundred times that quantity. One chemical may inhibit the production of a nerve-protecting enzyme, making it simpler work for a second material, with a history of attacking the nerve cells, to do its damage. Bacteria may also work on wastes to make them more lethal, and mixtures can transform stable compounds into ones that explode.

There was no way for the EPA to know how many potentially dangerous dumpsites were currently leaking, but one thing was clear: all possessed the capability of doing so. An EPA field study conducted in 1977 showed that of 50 landfills monitored by the agency, 47 were in some way discharging toxicants, including cyanide and mercury, into the ground. These landfills were chosen because they had no past history of problems. Clearly, no landfill could go unmonitored, for even one with the appearance of structural integrity could be wreaking enormous damage below ground.

The EPA's inventory of landfills was an arbitrary one at best, culled, according to one official, as much by "intuition" as from fact. What Jorling had received in response to his earnest directive to the regional offices was only a partial overview of the waste disposal process. Indeed, two regions, which included the highly industrialized states of Ohio, Indiana, and Illinois, did not even submit numerical estimates. Some offices had counted municipal garbage landfills, which frequently contain hazardous wastes, while others excluded them from their tallies. Few pertinent details were made available in the November assessment, and it was almost impossible to gauge the severity of conditions from what Jorling had received.

The EPA report stirred an immediate controversy, heightened by its appearance in the wake of Love Canal. There were a few complaints, for example from local officials in Missouri, that the EPA inventory of more than 30,000 dumps (or in the agency's vernacular, "environmental time-bombs") was unnecessarily frightening. But far more vocal and numerous were those who believed that the incomplete and fragmentary listing was merely a smokescreen to create the impression that EPA was on top of a situation that it had neglected. Congressman Albert Gore of the House oversight subcommittee described the EPA's assessment to the press as "inadequate" and "improper": "One gets the impression that they've really done something. But it doesn't appear they have. The material they released was ludicrous; it had no substance whatsoever. The way they described their statistics was very misleading—but a lot of newspaper reporters bought it. Jorling more or less ordered the regional people to look the other way." In reality, said Gore, America "has been pockmarked with thousands of cancer cesspools" while the agency had been studying pollution "from open stacks of cow manure in Vermont farms."

The shallow nature of EPA's comprehension of the problem was further exposed in February 1979 when an EPA consultant, Fred C. Hart Associates, reviewed the situation more closely and arrived at a figure of dumpsites almost twice as high as the first estimates. There were at least 51,000 of these sites, Hart figured, with 1,200 to 34,000 of them displaying potentially "significant problems." The Hart estimation, however, was also suspect. By its own account the report based some of its deductions on "scanty" information and unverifiable assumptions. For instance, waste sites utilized between 1930 and 1940 were not included, on the theory that chemical degradation would by now have rendered many of them innocuous. That may have been generally true, but there were documented cases of landfills from that period that were still posing hazards. It became clear that no one had a firm idea of the extent of the problem. It had grown beyond count.

The EPA files were not simply sketchy; they were also frequently outdated. For example, a chronology of incidents at a ten-block-long asbestos dump in Ambler, Pennsylvania, near Philadelphia, from which chemicals had migrated into a playground, ended abruptly at 1975, with indications that no more wastes were being added to the leaking spoils. In reality, the dumping was still going on, and state officials had not yet decided how to control the waste streams that were threatening a neighboring creek. An EPA log of a dumpsite in Basile, Louisiana, included these entries: "Ownership: Presumably the company. Type of business: Not specifically known. Types of hazardous wastes involved: Waste types are unknown. Size of business: Unknown. Condition of the facility: 55-gallon drums are reported to be floating in waste and water in an inspection made in *1973* [author's emphasis]." One call by a reporter to the state revealed that there were 2,000 barrels of caustics, acids, and other material and that the state was sufficiently distressed to have considered declaring it an official emergency in 1979.

More ominously, there were dozens of sites excluded from the EPA inventory that were of more serious concern than those on the listing. An EPA synopsis of trouble spots on which it had the most data failed to include many potentially grave and extensively documented cases in Texas, Louisiana, New Jersey, Michigan, and a host of other states. Among the missing was the "S" dump in Niagara Falls that was suspected of leaking into the city's water supply. In many cases, local newspaper libraries had more complete information than the nearest EPA office. The agency had listed 103 dumpsites on its tally as the best-documented problems, but it was obvious that the inventory was not only incomplete but grossly disproportionate. While Idaho, Oregon, Washington, and Alaska accounted for 16 of the 103 sites, only 4 were listed for New York and New Jersey when in fact these two states had more than 50 landfills that had been publicized as leakage problems. Congressman Gore cited another 16 dumps in his home state of Tennessee that did not make the list. The reason, EPA officials emphasized, was not that they were unimportant but that agency offices knew so little about them.

Thomas Jorling, for his part, acknowledged to reporters that the EPA's performance was "deserving of criticism." "As I pointed out during (House Commerce subcommittee) hearings, the agency has not been as aggressive as it should be in certain situations," he said. "But against that statement, we also have to balance the fact that we have very few resources to do anything, and we certainly have been focusing our program and its performance on getting a regulatory structure in place." What Jorling did not tell reporters was that his memorandum directing the regional waste surveys somewhat discouraged a detailed review of the crisis. "It is not expected that EPA make any effort to 'discover' sites (through field visits, substantial file searches, or other means) for which we do not currently have information," he wrote in his directive. "Rather, it is expected that EPA make estimates based on present and past industrial activity. . . ." He desired the data "only from your current regional files." Furthermore, Jorling did not want any conditions revealed of which the public was not already aware. "It is recognized that the development of this inventory will add national visibility to the incidents identified therein because the inventory will be shared with the Congress and will probably be requested by and made available to the public," he continued. "Because of this, incidents included in the inventory should be situations for which you have more than circumstantial information, the public (at least locally) is already aware, and publicly accessible information is already on file." To further discourage an in-depth review of the crisis, Jorling noted that state agencies should be contacted for information only if such an endeavor did not cause "undue burden." "There definitely is no requirement to go out and 'discover' hazardous situations," he repeated somewhat later, to make sure the regional directors had understood his message.

There were other indications that the EPA, far from simply being caught off guard, was evading the issue. In a 1974 report on hazardous waste sent to Congress from the agency itself, there were plain warnings that conditions were getting out of hand. "The management of the nation's hazardous residues—toxic chemical, biological, radioactive, flammable, and explosive wastes—is generally inadequate," said the introduction to that report. "Numerous case studies demonstrate that public health and welfare are unnecessarily threatened by the uncontrolled discharge of such waste materials into the environment."

Further along, the report stressed: "Inadequate management of hazardous wastes has the potential of causing adverse public health and environmental impacts. These impacts are directly attributable to the acute or chronic effects of the associated hazardous compound or combination of compounds, and production quantities and distribution."

But neither the cities, the counties, nor the agency had taken note of the warnings. No one had asked the industries to carefully register where they were placing their refuse and in what way. Now the effects of the waste, and of corporate and governmental irresponsibility, were appearing in the newspapers almost daily. The closing of drinking wells in Pennsylvania; a child rendered comatose in Nash, North Carolina, from playing with discarded pesticides; the disfigurement of two boys in Granite, Illinois, after falling through the ice of a small pond badly contaminated with leaking caustic soda: such incidents became common reading. It was not as if no one had foreseen the waste crisis. It was just that, for reasons of their own, the EPA and local authorities had chosen the role of quiet conspirator.

17

"GARBAGEGATE"

Amid charges that it had covered up ground pollution cases, the Environmental Protection Agency initiated a public campaign to demonstrate that it had coped with the problem as best it could. It was a fact that the hazardous waste management division was strapped for funding and manpower: of the agency's 10,900 employees, by early 1979 only 161 had been assigned to the division, and only a handful of these were working full-time at assessing new trouble spots. But it was also true that EPA's upper echelon had nurtured a climate in which the promulgation of new regulations to restrict wanton waste disposal was accompanied by years of infighting and red tape, and in which exposing new problems was treated nearly as an act of betrayal.

In February 1979, through its in-house publication, *EPA Journal*, the agency issued what it called a "waste alert"—a "major program" to increase public understanding of and involvement in the management of solid and hazardous wastes. Another goal, according to the EPA public information office, was to spur the nation's citizenry into provoking their local governing bodies to exert better control over the industrial rubbish hauled to the neighborhood dump. "The program," according to the journal, "will extend over the next several years and will involve citizens in all fifty states. EPA is being helped in the task by four nationally known organizations: the American Public Health Association, the Environmental Action Foundation, the League of Women Voters Education Fund, and the National Wildlife Federation." EPA bureaucrats began scheduling press conferences to show their newfound concern. Steffen Plehn, deputy assistant administrator for solid waste, now described the agency's survey of the dumpsite situation as "a real shocker, very frightening." Where only months before land dumping was an issue of low priority, suddenly the agency was giving Plehn's press conferences such titles as "Are Hazardous Waste Dumps 'Ticking Time-Bombs'?" or "Will There Be More Love Canals?"

Of course, the EPA already knew—had known for years—the answer to its own questions. A year before the proclamation of the "waste alert," Eckardt C. Beck, then regional director for New York and New Jersey, had described dumpsites in the agency's journal in terms identical with the ones headquarters was now using. "Even though some of these landfills have been closed down," Beck had written, "they may stand like ticking time-bombs." In like manner, more than eleven months before the state's designation of an emergency at Love Canal, the EPA had been repeatedly urged by the local congressman, John La Falce, to rectify the dangerous chemical oozing. La Falce, after receiving virtually no cooperation from headquarters, characterized the agency's response as "totally inadequate and abominable."

Again, there were dozens of other documented examples in which the agency was aware of hazardous circumstances but offered no aid to toxicant-plagued communities and promulgated no new policies to address the issue. By 1974, the agency files contained numerous accounts of wastes that had contaminated waterways, poisoned animals, and sickened men, women, and children.

For example, on December 7, 1971, at a chemical plant site in Fort Meade, Florida, a portion of a dike enclosing a waste pond ruptured, releasing an estimated 2 billion gallons of slime composed of phosphatic clays and insoluble halides into Whidden Creek. The waste flowed into the Peace River and an estuary of Charlotte Harbor, giving the surface a thick, milky appearance. Signs of life immediately diminished along the river, and surface fish activity was completely absent. The same was true for eight miles of Whidden Creek. Clam and crab gills were plugged with the white substance. That same year, in the Gulf of Mexico forty miles off the Texas coastline, a large number of chemical barrels turned up in shrimpers' nets, damaging the equipment and causing skin and eye irritations among the surprised crewmen. The containers came from two factories in Houston. Three years before, in Houston itself, cyanide, sulfide, and ammonia effluents had threatened the shrimp of Galveston Bay.

The EPA had also known about another dike, this one containing alkaline waste generated at a plant in Carbo, Virginia, which unleashed 400 acre-feet of fly ash into the Clinch River in 1967. The slug of contamination moved about a mile an hour downstream until it reached Norris Lake in Tennessee. In its wake, the poisonous trail left 216,200 lake fish dead and no sign of live food organisms for four miles of the river. Other incidents in Tennessee and Virginia and also in Missouri claimed substantial quantities of aquatic life, along with animals that fed at the shores. Shortly after PCB waste was deposited in the Waynesboro, Tennessee city dump, an oily slick surfaced in nearby Beech Creek, a brook so pure that, until the 1972 incident, it had been used as a source of drinking water. Apparently the waste, upon being unloaded at the dumpsite, was pushed into a spring that rose underneath the landfill and proceeded to the creek. Death was meted out to fish, crawfish, waterdogs, and raccoons for ten miles downstream. Uncounted hundreds of fish were poisoned by chlorine and ammonia that spilled into the Holston River from an industrial holding pond in Saltville, Virginia, and in Mosco Mills, Missouri, dumped pesticides killed 100,000 fish.

Meanwhile, toxic substances were finding their way up to farm mammals. Cattle died near the old Lowry Air Force Base bombing range outside Denver, which had been deeded to the city for a municipal dump; in Albuquerque, New Mexico, three children became seriously ill after eating meat from a pig that had ingested mercury-contaminated corn. On July 9, 1969, the owner of a farm in Patterson, Louisiana, noticed several pigs running out of a cane field squealing loudly, some of them in convulsions. Aldrin-tainted seed and several discarded containers had been dumped on the neighboring field. Eleven of the animals died.

Humans had also suffered directly. They had come in contact with hazardous waste in farm fields where it had been randomly scattered away from police view, or had been overcome by fumes upon opening drums trucked into landfills and solvent-recovery plants. That the handling of leftover industrial materials had created health hazards was fully evident: in the San Francisco Bay area, toll collectors became ill on a bridge across which alkyl lead wastes were being hauled.

One of the worst of the earlier incidents was in Ohio's Deerfield Township, and of this too the EPA had been completely aware. It involved a company, Summit National Services, which was allegedly in the business of treating and disposing of liquid industrial waste but which had simply allowed the residues to accumulate on its land until there was no way of handling them and fumes permeated a residential neighborhood. There were 100,000 gallons of waste oil on the premises and also substantial quantities of cyanide, paint sludges, sundry resins, and chlorinated hydrocarbons, including C-56 from Hooker's Michigan plant, in tens of thousands of 55-gallon barrels. Some of those containers were perched within a few feet of a highway drainage ditch that crossed the plant site, and dozens of drums were leaking onto the ground. A large cinder-block holding tank, poorly constructed at the joints, was allowing the seepage of an unknown toxic brew.

At least as early as September 29, 1976, and probably before, the regional EPA office had known about the conditions at Summit, but the agency's response was to avoid active intervention in the matter. The handling of Summit was revealed in internal agency memoranda, and it was disturbing. For example, on May 4, 1978, the agency's hazardous waste management division sent to Robert DuPrey, an official for EPA's Midwest Region V, a note that said:

> It has come to our attention that there may exist in Ohio an imminent hazard situation resulting from the storage of 100,000 barrels of hazardous waste. Specifically, our staff has learned that a company by the name of Summit National Services in Akron, Ohio, has an incineration facility that can handle a maximum of 500 gallons of burnable waste per hour. Preliminary calculations indicate that it would take about two years for this incinerator to burn the wastes that are presently on site if no further wastes are brought in. Because we feel that there may not be the financial resources available to this company to properly destroy these wastes, this facility

> may pose a potential imminent hazard. We request that Region V conduct a site investigation of the Summit National Services facility for possible imminent hazard action.

Another internal communication said that homeowners in the area had been complaining about taste and odors in their wellwater and that some had had to stop using their wells. Still another described the site as "deteriorating."

Instead of expressing equal concern and following the directive, DuPrey sent back a letter saying that it was up to the state to handle the problem.

> I do not intend to send any person from this office out to inspect the facility at this time, pending our assessment of the Ohio Environmental Protection Agency response. My *greatest concern* [author's emphasis] is the manner in which the term "imminent hazard" appears to have become loosely used by headquarters staff. As we are all aware, hazardous waste management facilities are *inherently* hazardous. Determination of imminent hazard is, in part, a legal matter and must in my view involve a risk of sufficient magnitude to warrant federal intrusion into an area that has historically been handled by the state and local sector. In the future, if you or your program staff have any questions concerning a facility in this region, please make the inquiry without identifying them as "potential imminent hazards." We are well aware that there are hazardous wastes in Region V. . . . In the future, I hope situations such as this can be brought to the attention of the regional office in a more rational manner.

When regional administrators were not resisting the efforts of the hazardous waste management division in

Washington, the highest officials at headquarters took up the task, discouraging the few investigators looking for danger areas from aggressively pursuing their search. Some of these investigators grew distraught enough at these restrictions to begin publicly attacking the agency. The most vocal was Hugh Kaufman, a free-wheeling thirty-seven-year-old whose sweater vests, rimmed glasses, and neat, short-cropped brown hair—reminiscent, in his own words, of a soldier in the Republican political cadre—belied his disposition to whistle-blowing and public protest. When I met Kaufman, who manages the hazardous waste assessment program, he was sitting in his small office in Washington with his loafers propped on his desk near a small American flag, sucking small quantities of tobacco tucked into the base of his left cheek. "What we have to do is force the system to help the people out there," he told me. "Costle and Jorling and the White House are interested in covering up imminent hazard situations because they don't want to pay for the ultimate remedies. All they're concerned about is the political institution—protecting that. So they lie under oath and try to stop us from going out there and finding sites. It's easy to sit up here and forget about the people who drink pesticides in their water." Kaufman said that on June 16, 1978, he had received a memorandum from his boss instructing him to "stop looking for imminent hazards." The policy, Kaufman charged, was one of benign neglect, in the interest of the nation's economy.

On October 30, 1978, several months before I spoke with him, Kaufman had appeared before the House Subcommittee on Oversight and Investigations with his charges. His testimony was as follows:

> I prepared and circulated to management a policy recommendation memorandum and began to search out

hazardous waste facilities that posed a threat to the public health and environment. The first case we found was at Seymour Recycling facility in Seymour, Indiana. As a result of our efforts the EPA regional office was forced to pressure the state to take action against this facility. After working on the Seymour case, we began to uncover other cases which posed a potential threat. However, trying to get the appropriate EPA regional offices to take action against these facilities proved to be an almost insurmountable obstacle. In one case involving the Summit National Services facilities in Deerfield, Ohio, the EPA regional office refused even to visit the facility or to let headquarters personnel visit the facility. This regional position was supported by our management and therefore no visit was made to this facility. When I discussed this case with the regional officials I was told that they have in their files other cases that are even worse than Summit and stated that they didn't plan to investigate any of them, let alone take action against them. They also told us to stay out of their region. On September 4, 1978, this case appeared on the front page of the *Washington Post*. The next day the regional office notified us that their policy on the case had changed and that they now planned to take action. I should add that EPA headquarters enforcement first learned of this case by reading the *Washington Post* story, not from the region. When I reported this to my management, no action was taken. In another case involving the Velsicol disposal site in Hardeman County, Tennessee, the EPA regional office told us to stay out of the case and when we raised the issue to management we were also turned off the case.

Most disconcerting to Kaufman was that there were no EPA guidelines "to alert the public to potential danger; in fact, the memo further instructed the regions *not* to find new problem sites because they might be required to provide this information to Congress and the public."

A few days after his testimony, Kaufman entered his cubbyhole office to find a picture of a pig sitting on his desk. A couple of days later, there was a piece of stale cheese in the same spot. Steffen Plehn had recently asked Kaufman to "join the team," said Kaufman, but he "had hit my moral ceiling. There were people out there being poisoned and the EPA was covering it up, or going to a strategy of a 'limited hang-out,' like they did with the waste site list. Here they spend more money putting out the journal issue of the 'Hazardous Alert' than on looking for more sites, and here we have something that makes nuclear waste look like Mickey Mouse. With their list of hazardous sites, what the EPA wanted to do was demonstrate that the costs are so exorbitant to clean them up that their policy of not looking for more is the correct one. Even now, I'm not allowed to look for new sites. And I wasn't even consulted on the list of dangerous sites, even though that's supposed to be my job. What we have is another scandal on our hands: Garbagegate." The June 16 memorandum Kaufman referred to had been sent from Plehn, with the specific instructions that the division "put a hold on all imminent hazard efforts." The reason for that, Plehn explained to me, was that jurisdiction for such matters was being defined under the agency's enforcement division while his unit was culling a "data base." The problem, according to Plehn, was simply bureaucratic. Plehn further explained that it was "a deliberate programmatic choice" so the agency could place primary emphasis on preparing regulations for the Resource Conservation and Recovery Act of 1976 (RCRA), which when fully implemented will provide federal standards on toxic waste disposal.

Specifically, RCRA, which was signed into law by Gerald Ford on October 21, 1976, mandates the design of a system to eliminate open dumping, to inventory waste sites, to regulate how landfills are constructed and monitored, to require private funding for long-term care of the dumpsites, to send grants to rural communities to improve solid waste management programs, and to institute a permit, or "manifest," system whereby chemical material is kept track of in record form "from cradle to grave." Persons owning or operating facilities for the treatment of hazardous substances will be required to obtain permits that list composition, quantities, and the rate of disposal of such rubbish, as well as dump locations. (Exempt from this requirement are farmers, retailers, and any concern producing less than 220 pounds of hazardous waste per month, a statute that raised controversy among the environmentalists, who pointed out that even small quantities have been known to pose health hazards.) Within five years of the dump inventory, to be carried out by the states and localities, all open sites will be closed or upgraded according to new strictures devised at EPA. RCRA also created, within the agency, the Office of Solid Waste Management Programs, and authorized EPA to seek court declarations of dangerous situations as "imminent hazards" and to force industries to pay the price of cleaning them up.

There were many defects in RCRA, both in its design and in the fashion in which EPA was implementing the new legislation. One inherent fault was that it dealt mainly with active or future dumps, ignoring abandoned sites no longer under the liability of the generating company or waste hauler. "There is no disagreement that inactive sites are a public menace," protested Leslie Dach, a scientist for the Environmental Defense Fund. "Yet protection of the public from the hazards associated with these facilities has been neglected by EPA. The main reason they don't want to look at them is the bill." There was also the problem of timing. When the Comptroller General's Office reviewed RCRA in 1978, it predicted that "the improvements mandated will not be accomplished within the legislative time-frames." Indeed, where by law EPA was supposed to issue specific regulations within eighteen months of Ford's signature, partial implementation now is not expected to come before 1980 at the earliest, with the law not fully operative for perhaps years. Meanwhile, waste haulers and small landfill owners who would not be able to afford to operate under the RCRA restrictions were hurriedly piling in waste to beat the deadline, thereby creating more problems for future generations to deal with. Because of the delays, at least 160 billion pounds of additional hazardous waste will find its way into substandard settings.

The first and perhaps most important regulations EPA had to devise were those in Section 3001 of RCRA, concerning the definition of those wastes to be considered "hazardous." It was the design of this definition that caused the most internal dissension and helped delay implementation. Ignitable, corrosive, reactive, or toxic chemicals were to be included, as determined by a series of laboratory tests. The required tests, detailed in a March 1978 draft of the regulation, were extensive ones, covering a broad range of effects that included the capability of compounds to cause mutagenesis (genetic changes) in living organisms. The draft went through a "working group," then through a "steering committee" and a "red-border review," and finally on to the administrator. The process took so long the EPA was sued by the state of Illinois and the Environmental Defense Fund for delinquency in attacking what they saw as an immediate crisis.

By the time the Section 3001 regulations were printed in the Federal Register (as well as Section 3002, which concerned testing for toxicants in their wastes on the part of generators and haulers), there had been a substantial reduction in the range of compounds considered hazardous and in the obligations of industry to fund such testing. The agency had decided to cut down the scope of the regulations, ostensibly because there were not enough test results available for many compounds and because complete testing of every waste stream would be prohibitively expensive for government and industry alike. No longer were mutagenic chemicals necessarily listed under the heading of "hazardous waste," and originally missing from the section were such notorious toxicants as C-56, Mirex, Kepone (which destroyed fishing in the James River in Virginia), one waste known to contain dioxin, residues from the manufacture of endrin and heptachlor (two suspected carcinogens), and a dye intermediate, o-anisidine, which had caused tumors in more than 90 percent of the animals tested.

The EDF responded to the list in the Federal Register with predictable fury: "As currently proposed in the Section 3001 regulations, a waste is considered a toxic hazardous waste only if it contains above a specified amount of a substance for which there is a National Interim Primary Drinking Water Standard, or if the waste or the process generating it are specifically listed by EPA," it said. "Without question, this approach exempts significant amounts of hazardous waste. It makes a mockery out of Congress' intentions and the public's need for protection." Many pesticides, a number of which were carcinogenic, fetus-deforming, or mutagenic, were excluded from the list. At a high-level session, officials had discussed excluding oil and gas drilling muds from the list; the internal notes disclosed that one administrator advised them:

"Don't take on oil and gas industry at outset." Kaufman said the electrical industry, oil production, and manufacturing chemists' lobbying groups had all applied pressure to ease the strictures, with apparent success. "In short," said the EDF, "the list is based more on historical accident than thorough science, and, therefore, is hardly a complete listing of all processes generating hazardous waste. Even if the EPA adopts our suggestions of lengthening the list of specific chemicals and the list of processes, wastes hazardous to the public will still be exempt from the RCRA program. Sole reliance on lists relegates the RCRA program to those wastes or waste constitutents about which a substantial amount of toxicological information already exists. Because the number of chemicals tested for toxicity is so small, the amount of hazardous material exempted by the listing approach may be large."

The shortening of the hazardous-waste list reportedly had its origins in a June 15, 1978 meeting held by John P. Lehman, division director for hazardous waste management, for his staff. According to notes taken in the June meeting, Lehman told his employees that President Carter's directive to reduce the federal budget "is putting a tremendous squeeze and pressure on the agency as to our ability to carry out these new programs, not counting old programs." The Proposition 13 tax revolt in California, he told the staff, was likewise causing pressure on the state to spend less on regulatory programs. As a result, of course, the RCRA regulations were to be more limited in nature.

This discussion and the subsequent actions brought a quick and bitter response from William Sanjour, who until he loudly complained about the handling of RCRA had been Kaufman's immediate superior at EPA. Sanjour protested, in a March 15, 1979 memorandum to Lehman, that the new regulations would allow "80 percent of the waste from the manufacture of pesticides" to remain uncontrolled. "I would also like to point out that earlier drafts of the regulations had test protocols which would have prevented this from happening," he said. Sanjour claimed that it was his understanding from the meeting that, in order to save the nation's economy and keep industry happy, full implementation of RCRA regulations was about to be delayed for three years and maybe more. He leveled many of his complaints at Gary Dietrich, who was on Jorling's immediate staff at the time. Sanjour wrote in a notation to the record:

> I pointed out to Dietrich that most people who are being exposed to poisoning and other danger of hazardous waste would continue to be exposed during this interim period of over three years. I asked him if the agency could withstand the political pressure from the people living next to these facilities which the law requires us to protect them from. Will the public stand for us sacrificing their health because we have some over-all concept of where we want to go and we don't want to pain industry too much while we get there? He said he was willing to "risk" it. I commented the public health is what is being risked. I asked him how he would explain this policy to people breathing the hazardous fumes and drinking the poisoned water. He replied, "I can be facetious and say you've been doing it for the last ten years—so what?" And he laughed. I told him I could not go along with his proposals and I did not think he was acting in good faith.

Sanjour further charged that he and others were "told to avoid regulating hazardous waste from the oil and gas industry, electric power companies, and other large industries" and "required to write public documents which we knew were misleading. The press and Congress and the public were given misinformation while accurate information was suppressed." In the meantime, he instructed his staff to take oil and gas muds off a "special waste" category that would have exempted them from most standards for disposal. "I informed my management of this and I instructed my staff to remove drilling muds from the list of 'special waste.' A few days later I was removed from my position and drilling muds were reinstated as a 'special waste.' "

Because under the RCRA mandate the agency had been charged with developing and promulgating "criteria for identifying the characteristics of hazardous waste," Sanjour labeled his superiors' regulation cutbacks "immoral and illegal." In the Register, the EPA had explained that its scope was based on available testing methods and toxicological data already in existence that were applicable to waste. But Sanjour interpreted that as merely an excuse contrived for ulterior motives, since in his discussions on the regulations he had not previously heard of the test protocol as the key issue. "One must conclude from all this that the test protocol story in the Federal Register was invented after the decision to drop the hazardous waste characteristics had been made and that the real motivation for the decision was an attempt to reduce the cost to industry in response to the President's policy of fighting inflation," Sanjour claimed. "The lies—you can build mountains on them. The whole organization has been devoted to a cover-up. Their actions are despicable. I resisted. I fought tooth and nail. Plehn and Jorling got very mad. They knew what they were doing was illegal and they did not want me there documenting it. So I've been detailed to a meaningless job, as far from hazardous waste as they can get me."

Despite the occasional tone of Kaufman, Sanjour, and other internal dissidents, it was not as if the EPA brass had conducted any conspiracy to endanger the health of citizens living near chemical receptacles. Plehn was forthright in admitting that economics was a heavy consideration. At a press conference in Boston, when asked if the regulations had been watered down too much, he told reporters that environmental needs must be balanced "with other social and economic needs." For example, the requirement that companies monitor their facilities for forty years had been reduced to twenty years as a result of an estimated cost, just for observing this regulation, of $800 million a year—most than four times what industry was currently expending on treatment of hazardous wastes. Those figures were debatable—current costs seemed to have been underestimated—but Plehn's statement showed clearly that EPA's main concern was the presidential budget. The simple reality was that the office of EPA administrator was very nearly a cabinet position, serving largely at the pleasure of the president and the Office of Management and Budget. The agency's own budget acted as a severe constraint as well. For fiscal year 1979, it had been appropriated $1.2 billion; only $25.2 million went to the branch overseeing hazardous waste management, and of that, substantially more than half the dollars were going into efforts at implementing RCRA. Land pollution efforts were granted only 3 percent of the agency's resources, a tenth of which went each toward preserving water and air.

Had the entire annual budget for EPA been sunk into the division of hazardous waste management, there would still have been huge shortcomings. The agency's consultant, Fred C. Hart Associates, estimated that even to contain temporarily one non-nuclear chemical landfill would cost $3.6 million, while a permanent solution would rise to $25.9 million. This meant that if there were, as the EPA originally estimated, 1,100 chemical dumpsites constituting "significant problem sites" (excluding landfills containing radioactive wastes), it would cost more than $26 billion to rectify the immediate situation on a long-term basis. Hart, however, calculated that there were more like 1,703 significant depositories (not including nuclear wastes), which would have put the cost figure at about $44 billion. Not all of these would have to be paid for by the taxpayer. It was estimated that for perhaps as many as half of them, there was the chance that specific industries could afford the costs; and at 26 percent of the sites, some form of remedial action had already been taken. But at best the cost to government threatened to be far beyond any possible allocation.

With these considerations in mind, President Carter, who had called the presence of toxic substances in the environment "one of the grimmest discoveries of the modern era," asked Congress on June 13, 1979, to create legislation that would impose fees on industries to clean up both hazardous waste and oil spills. His request, channeled through EPA administrator Costle, called for a $1.6 billion fund over the next four years, 80 percent of it raised through fees on raw industrial materials and the rest paid by the federal and state governments. The proposal also included provisions for the federal government to clean up and "mitigate" pollution where the responsible party does not take immediate action or cannot be immediately identified, with the government later taking court action to recover these costs, and requested stricter standards for the handling of hazardous substances. Moreover, Costle said the president wanted $46 million added to the agency's 1980 budget to deal with emergencies created by abandoned sites—a paltry sum over all, but a significant step nonetheless. Industrial fees would be up to 3 cents a barrel for oil and petroleum, a half-cent a pound for petrochemical producers, and $2 a ton for inorganic chemicals and heavy metals. Jorling explained that this would result, for example, in just a four-hundredths of a cent increase in the price of gasoline at the pump. "Both industry and consumers have financially benefitted from cheap and unsafe disposal practices in the past," commented Costle, "and therefore both should share in the remedies we must now pursue." The Chemical Manufacturers Association's president, Robert A. Roland, felt differently. "The bill unfairly singles out the chemical and related industries to bear a disproportionate burden of the cleanup costs," he said, displaying the type of resistance the agency constantly encountered. "In so doing it fails to adequately reflect society's responsibility for resolving a problem which everyone has helped to create and for whose solution everyone should help pay."

The shock of these measures could have been eased had the federal government more gradually, and much earlier, responded to the solid waste dilemma. In the 93rd Congress, a Hazardous Waste Management Act had been proposed but had not received approval. In 1975, during the 94th Congress, it once again became a serious topic of conversation for the House Subcommittee in Transportation and Commerce, which had called in John R. Quarles, Jr., then deputy administrator of the EPA, to answer some questions. When Quarles told the subcommittee chairman, Fred Rooney, that the EPA did not plan to resubmit the legislation "in light of President Ford's decision to hold the line against all new spending programs in order to fight inflation and keep the budget in control," the exchange grew a bit testy.

MR. ROONEY: One of the greatest problems facing this nation, and we heard witness after witness testify, is the solid waste problem facing this country today. And we are spending billions of dollars on unnecessary spending in other areas and why should we get so chintzy in an area that means so much to the health and welfare of this nation?

MR. QUARLES: I think that is a broad brush type of question.

MR. ROONEY: It is not a broad brush type of question. What you said is shocking. Here we are trying to solve a problem and you say just because we are trying to battle inflation we can't spend another nickel.

Shortly after, Rooney's frustration became even more obvious:

> It seems to me, Mr. Quarles, and I am not quarreling with your response, but it seems to me every time, not only this administration but every administration I served under for the past 12 years, every time there is a cutback, it is in one of the three areas of health, education, or welfare, and I think we can look to other areas in which to cut back, but certainly not in this area.

At the time, in the words of EPA, there was not a single state "adequately" controlling disposal of wastes of a toxic nature. But this was allowed to continue for years, at an indeterminable cost to the lives and health of Americans.

In other arenas, interestingly enough, the EPA assumed a different stance. Douglas Costle wrote in the January 1979 issue of his agency's journal:

> None of us can be unconcerned that prices continue to rise, that Americans are again jittery about having to pay

more and more for essentials—food, shelter, and clothing. As the President leads the attack on inflation, we at EPA must be concerned about whether the environmental program contributes to the inflation rate. Some say the root of inflation is in the government's monetary and fiscal policies. Others emphasize excessive wage settlements and price increases that don't reflect increased product value. But there are other reasons as well, including possibly the effect of costs imposed by regulations. So we must continually evaluate our actions to be sure that they are not unduly inflationary. As measured by standard yardsticks, such as the Consumer Price Index, EPA's programs do contribute modestly to inflation. Our most recent analysis, done by the respected firm of Data Resources, Inc., estimates that EPA's air and water pollution control programs will add an average of 0.3 percentage points annually to the Consumer Price Index from 1970 through 1986. Thus, if the Index were to increase by 6.0 percent in a particular year without pollution controls, it might increase by 6.3 percent with them. The results of the Data Resources analysis are in step with earlier studies done for EPA and the Council on Environmental Quality. All indicate that while the impact of pollution control on the Consumer Price Index is noticeable, any conceivable change in current regulations wouldn't substantially alter the nation's underlying inflation rate. . . . We can think of cases where environmental protection is (or would have been) clearly worth the investment. For example, the Kepone contamination of the James River has shut down the fish and shellfish industry in the area—probably for decades—because it would cost billions of dollars to clean the river bottom of the contamination. Preventing the problem in the first place surely would have been less costly.

His deputy, Barbara Blum, added that "somehow, somewhere, there has developed the myth that it is inappropriate for us regulators to be interested in things like free enterprise, inflation, and economic growth. That myth has been supported by another: that economic growth and environmental protection are fundamentally at odds. These myths deserve to be debunked."

However hypocritical the rhetoric of EPA officialdom, it accurately reflected the mood of Americans. In a 1978 poll by Resources for the Future, by a 3-to-1 margin people chose to pay higher prices for air and water protection if that was what was necessary, and further analysis showed that among those who felt taxes to be "very unreasonable," the highest percentage still believed the environment must be improved "regardless of cost." Clearly, the tax revolt had not distorted the public's view of what was important. It was the EPA that often did that.

For all of Costle's and the president's lip service and tentative programs and surveys, the hard decisions were still being made almost entirely in terms of both cents and megadollars, with human health a secondary consideration. Such was the opinion of a House commerce subcommittee chaired by Congressman Robert Eckhardt, which concluded in October 1979, after months of testimony, that the EPA had taken an "inexcusably" long time in regulating hazardous waste, so that there was now an "imminent" threat to public health that "can not be overstated." It would take years, said the subcommittee, to make up for the lost time.

But like personnel in other large governmental institutions, those who worked for EPA were subtly induced to put their personal career fortunes above public service, causing further delays. As an Agriculture Department employee wrote in the *New Republic*, the key to advancement in a department or agency was to "please your boss, cover your ass, and always, always be cautious. Patience was the greatest virtue. The way to get ahead was not to outshine everyone else, but to do precisely what your superiors wanted, prove your loyalty, and get to know everything you could about the bureaucracy's inner workings." These same attitudes were also common at state health departments, at county seats, and in city halls throughout the nation. Bureaucrats frequently protected themselves rather than the victims of toxic wastes.

The EPA's headquarters, known as Waterside Mall, about two miles south of the Lincoln Memorial, was an apt representation of the attitudes and operations of the agency. The hallways were long and canyonesque, their walls a hypnotizing off-white, and there was such a maze of hidden corridors and cubicles, of uncoordinated floor levels and elevators, that even those who had worked there for years were prone to disorientation. Visitors were given a map, but this did not always help. In the hazardous waste management division, there were stacks of duplicated paperwork and rows of cabinets and phones, but the files on dumpsites were so sparse that they occupied just three cabinet drawers and lacked vital data on a number of the most troubled spots. Around them, the staffers sat with their noses buried in complex regulation proposals, or propped their feet on their desks to relieve the boredom of the afternoon. The exterior was basically two towers with a low-slung shopping center and office space in between, a canopy of cement crutched on pillars, with endless rows of rectangular windows. The towers had a special symbolic significance: they had the appearance of aloofness and inaccessibility, and like the ears and eyes of the administrators, they constituted a wall of stone.

EPILOGUE: THE ROAD BACK

During the summer of 1979, the New York State government took a startling new tack in its dealings with toxic waste problems, one that dramatically bared the inner thoughts of bureaucrats responding to environmental threats. In a sterile tower that looks down upon the gargantuan Albany Mall complex, researchers worked on a draft study for the Department of Health concerning contaminated groundwater in general and the probability of chemical substances causing cancer in particular. There was a well-founded apprehension within the department over how the public would respond to the study if it was released, for it was, in essence, an attempt to estimate and to articulate the financial value of saving people from exposure to toxic wastes. The state, having just witnessed the tragedy of Love Canal, and faced now with dozens of widespread well-contamination cases, had decided it was time to determine the cash value of salvaging human life and health from toxic waste jeopardy.

A key section of the 130-page draft, subtitled "The Benefit of Reducing Risk," contained six empirical estimates of what a life means financially to society. "Estimates range from $49,226 to $1 million with most values between $200,000 and $300,000," it said coolly. "These estimates will be used later to describe the benefit of reducing the risk of death." The department researchers took into consideration such factors as an individual's income, his productivity, and what it would cost to treat his ailment in order to determine whether the state's monies would be wisely invested in reducing the possibility of cancer. "Despite justifiable moral and philosophical objections," it noted, "this dilemma must be confronted if economic assessments are to be made."

With that, the state scientists analyzed various models of determining an individual's worth and the cost of installing water purification systems to reduce health hazards. They arrived at two formulas: $B = (r_1/70 \times v$, where B stands for the per capita benefit in dollars a year, v is the assumed economic value of saving a life, r_1 is the per capita lifetime risk estimate, and 70 the average life expectancy; and $r_1 = .95\ C \times V \times r$, where C is the amount of contamination, V the per capita daily water intake, and r the lifetime risk for an oral dose. The value of preventing one death, as determined in these formulas, ranged from $100,000 to $1 million. As a basis for making these estimates, it was suggested that the researcher should "assess the worth of an individual's present production. Further production is discounted because it is generally believed that future production is less than present production." A weakness of the specific methodology, admitted the report, was that it would "undervalue lives of housewives, elderly, unemployed, and underemployed," and provide "no allowance . . . for social values or the utility of life to an individual." On a more sensitive side, the report noted that while it was unlikely that one or two dozen extra deaths from toxic exposure in the entire state would cause public concern, "these events certainly would be noticed by [those persons] directly affected, their friends, relatives, and business associates." That burst of warmth aside, the report went on to conclude that whenever trichloroethylene was greater than 50 parts per billion in drinking water, based on a risk of one extra death in a population of 100,000 and an assumed economic life value of $500,000, it would be worthwhile to install aeration treatment to decrease contaminant levels in a system serving 10 million gallons a day, but that "treatment of smaller systems cannot be justified based on these data." Therefore individuals who owned their own wells or who were attached to small municipal systems would not, based on economic considerations, be provided the proper protection.

The state's calculations were flawed in many respects. On technical grounds alone, the study had serious shortcomings. There were no numerical expressions to indicate possible teratogenic, mutagenic, central nervous system, liver, or blood effects, nor did it assess noncancerous diseases. The study was based on the probability of cancer alone and even in this it was deficient. By the draft's own admission, "the exact mechanism of carcinogenesis is unknown. The existence or non-existence of a threshold level has never been proven. If one cellular change can lead to a malignant transformation and a lethal cancer, there will be no threshold level." And certainly no safe, or "acceptable," one. Whether a resident is drinking 1 part per billion of carbon tetrachloride, or 1 part per million, is immaterial. Both may produce cancer. Because there can be no experiments with human subjects, a chemical's toxicity can be gauged only by its observed effects on a laboratory animal. Such a method is not always accurate. There are indications, said the state, "that a substance will be less toxic in smaller animals than in man." Indeed, thalidomide had caused birth defects in women at one-hundredth the dose observed in a scientist's rats, and certain of the aromatic amines induced cancer in humans but not in treated mice. Moreover, the state's calculations did not take into account the reactions different chemicals have on each other when blended in a common soup. Often such an intermingling, as I have stressed, greatly increases the toxic effects. When certain organic phosphate pesticides and chlorinated hydrocarbons enter the body at the same time, the damage they wreak is ten times what they would inflict individually, for the chlorinated hydrocarbons damage the liver in a way that prevents the enzyme cholinesterase from properly protecting the nerve tissues, leaving them open for attack by those organophosphates that specialize in causing damage.

But New York's intriguing report is best criticized on its moral and philosophical grounds. Is one life truly worth more than another, and can a monetary figure be attached to the difference? Does the Constitution not provide for equal protection under the law?

Whatever its merits, such thinking had long been common in the chemical industry. Spokesmen there charged that a "chemophobia" was pervading the nation, and that the risk of chemical exposure is a necessary one, comparable to that of driving a car or riding a plane. They neglected to mention that those latter risks are voluntarily undertaken, while an individual does not have the same choice in drinking water or breathing air. Unembarrassed by hyperbole, the chemical manufacturers, in slick advertising campaigns, implied that life itself would not be possible without the finer medicines, more hardened plastics, and more potent pesticides that come forth from their steaming vats.

For years the industry had shielded the true dangers of its chemical processes and wastes from public purview and actively pressured politicians from full inquiry. All the while it was fighting stricter regulations, it knew that more cautious handling of its waste materials would not destroy the industry but only cut into its often exorbitant profits. The manufacturers not only downplayed the dangers but on occasion actively covered them up or kept them from entering the newspapers. The Manufacturing Chemists Association once maintained silence about its knowledge of the dangers associated with vinyl chloride; the Allied Chemical Company for years reportedly kept from public scrutiny the knowledge that its pesticide Kepone caused liver and central nervous system disorders. Even the august Stanford Research Institute had been accused of falsifying tumor studies for the sake of the Shell Chemical Company. In 1978 it was alleged that Stanford had cut tumors out of test animals and had not mentioned them in certification reports to the government, a charge it denies.

Industry's blind obsession with economics has often been abetted by government. In February 1979, a controversy developed between certain officials at the Environmental Protection Agency and White House economic advisers Alfred Kahn and Charles L. Schultze: the administration, despite President Carter's generous campaign pledges to protect the environment, pressed for relaxation of water pollution regulations, with the goal of reducing control costs to industry. When reports grew of resistance within the EPA to this policy, the president's press secretary, Jody Powell, informed news reporters that those dissatisfied "should be aware that their resignations will be gladly accepted at the earliest opportunity and should not be hesitant at all in offering them." That arrogant message provoked a counterattack by Senator Edmund Muskie, who, unlike those at the White House, had spoken with victims of Love Canal. Quite accurately, he accused White House "inflation fighters" of "treating public health as inflationary" and acting contrary to the intent of Congress in its passing the regulations to begin with. In August came another salvo at public officials in the form of a seventy-eight-page report issued by a Rockland County grand jury which, having investigated cases of indiscriminate dumping, felt the need to sharply criticize government "at all levels" for the fashion in which it has dealt with "flagrant and widespread" hazardous waste practices; this has fostered "actual and potential criminality and profiteering," the report said, adding, "The evidence indicates the response of federal, state, and local governments to the problems posed by hazardous waste has been characterized by ignorance, neglect, laxity, and fractionalization of responsibility."

In such a milieu, it is not surprising that the United States finds itself today in the throes of what can only be termed a cancer epidemic. Cancer now accounts for nearly 20 percent of the deaths each year, second only to heart disease. About 55 million people now alive have contracted or eventually will contract cancer. Nor can increased longevity alone explain why more people are cancer victims, since so many children suddenly are succumbing to the disease. It seems too coincidental that the increase in cancer rates has so closely paralleled the increase in rates of chemical waste production, especially those of the solvents and halogenated hydrocarbons.

We can, as a society persist in our self-destruction. Or, we can begin now to reduce the slaughter. But for it to have an effect, we must do it soon. Some of the measures necessary to halt the contamination of our land will be relatively painless to achieve, and some are already mandated, in varying degrees, in laws such as the Resource Conservation and Recovery Act of 1976. Some will be more drastic and will infringe on the extravagant life-styles we have nurtured. But all are vital.

The immediate steps to be taken are not dramatic ones. They include keeping landfills away from major aquifers, rivers, and lakes, and especially from residential clusters. It would be prudent to set a standard location at fifteen miles or more from a central source of drinking water to ensure against long-term leaching. Ideally, landfills should be situated in regions where evaporation exceeds precipitation. Their construction should include thick, solid clay bases; improved synthetic liners on the sides and bottom; leachate withdrawal stations that run the length of the landfill; interception trenches; and on top, another coating of plastic, itself covered by a clay cap planted over with grass. Chemical drums must be well marked and carefully placed, and they should be implanted in grid fashion so that various types of chemicals do not intermix (and may even one day be excavated for recycling when the technology arrives). Those who operate landfills should be required to attend training sessions on the toxicity of chemicals and to pass a written examination. Operating firms must be forced to set aside large sums of money in performance bonds and trusts so that if they go bankrupt or simply abandon their pits, there will be the money to close the site and monitor it for leachate. Chemical manifests recording wastes from the time they leave the factory to their ultimate disposal should be required and monitored by newly created state agencies with the manpower to make regular random checks that toxicants are disposed of only in secured landfills. Finally, serious criminal and civil sanctions should be imposed on those who violate these regulations, and landfill operators should be required to install air-monitoring devices so that governmental agencies can check for evaporation into the already chemically overburdened atmosphere.

But these are temporary measures. No landfill can be

considered forever safe, so in the long run their very existence is not acceptable. The next step would be to ban the discharge of any chemicals into the ground except in unusual cases. If such a policy, however startling and unreasonable it may now seem, is not effected before the beginning of the next century, an intolerable number of acres will be rendered both useless and dangerous. In a nation that depends so heavily on the productive utilization of its soil, the dwindling of such a resource is intolerable. Should we continue burying our wastes, too many more fields will be despoiled, creeks barricaded, and homes boarded up and deserted.

In the years to come, the amount of toxic wastes destined for ground disposal will increase significantly. More chemicals are being manufactured, and waste sludges amounting to nearly 8 million dry tons annually are arriving from our new wastewater treatment plants. Other wastes once dumped into the ocean off barges but now restricted from that disposal method are being sent to the land. Because it is nearly impossible to remove a landfill once it is there, disposal provides no solution; the answer is disassemblage or direct destruction of waste products. Glimmerings of the next generation of sophisticated waste disposal systems have already appeared: a few firms have begun to separate waste materials through sedimentation, flocculation, filtration, and membrane osmosis; to collect and concentrate them through a process known as ion exchange; to reduce their toxicity by oxidation or by nurturing micro-organisms to feed on the wastes; and to render them into coke or activated charcoal by heating them in an oxygen-free atmosphere, the process known as pyrolysis. Even these advanced procedures do not totally eliminate toxic residues. A major research effort must be instituted, by government and industry, so that in the end toxicants will have been pulled apart and returned to the environment as simple, benign molecules. By the use of ozone or oxygen, or ultraviolet light and other as yet unseen forms of irradiation or chemical reactants, the chemicals might be completely dismantled. Small amounts of residue that remain could be solidified in rocklike form and set in secured landfills so that they do not leach. One way of doing that, suggested by a Pennsylvania waste firm, might be to add cement and aluminum silicate to these ultimate dregs, trapping the chemicals in a polymer lattice as impermeable as clay.

While such research efforts are proceeding, there is a current need to reduce landfill-destined wastes by constructing effective regional incineration centers. For some compounds, high-temperature incineration is the only acceptable method of destruction. If properly managed and stringently monitored, and kept out of the hands of irresponsible corporations, such processes minimize the potential of adverse chronic exposure. For years, waste incineration has been employed by companies such as 3M Corporation, Dow Chemical Company, and the Kodak Company. While expensive to operate, they have been quite successful. Most use a combination of rotary kilns (horizontal brick-lined furnaces that turn as heat is directed at the inflowing chemicals) and secondary combustion chambers. Both liquid and solid waste can be burned in the units at temperatures ranging from 1,100 to 2,400 degrees Fahrenheit, and the residues deposited in a quenching chamber to be cooled to the point where vapors condense. A water scrubber cleans flue gases of fly ash and the precipitated vapors, and the solution is sent off to a wastewater treatment plant while the ash heads for a landfill. At a facility in Van Nuys, California, an experimental incinerator set at 1,348 degrees Centigrade destroyed the dense, dark brown fluid of C-56 and other organic compounds with 99.97 percent efficiency. While the cost was high—$487.59 per metric ton—no evidence of C-56 was found in the residues. Much of the incineration ash was composed of inert chemicals, salts, and heavy metals (which can still create land pollution problems) and the air emissions contained hydrochloric acid as well as carbon dioxide and water. These heating units should be placed a good distance from residences and monitored constantly. Perhaps they would best be operated on barges out at sea, for one puff of a potent organochlorine compound escaping by error from the stack would cause more serious contamination than the leakiest landfill. Wherever they are located, incinerators should be regarded only as interim solutions to waste disposal.

Those waste pits already in the ground, if their contents cannot be safely incinerated or otherwise treated, must be contained and their environs thoroughly investigated for contamination. Special attention must also be paid to those thousands of nasty open waste lagoons on back plant properties. All should be inventoried, catalogued, monitored, and publicly listed. Aerial surveys using infrared or ultraviolet filters have been found to be effective in detecting migration of leachate; if the dump is escaping its bounds, a drainage program must be implemented to capture the liquids before they reach a major aquifer, and persons living nearby should be tested for liver or pancreatic disease.

In the meantime, we need more research on ways to remove toxic substances from human bodies once they have lodged in the cells. For example, some success has been achieved in extracting metals from the body by giving the victim dosages of what are termed "mixed ligand chelate agents," which combine with the metals and carry them to excretion. In the same way, a cholesterol-reducing drug, cholestyramine, has been found to speed the removal of Kepone from human bodies.

Rather than developing increasingly elaborate procedures for the destruction of toxic wastes, we can minimize production of such substances in the first place. This alternative is one that would include the curtailment of certain chemical processes and, if necessary, an outright ban on the manufacture of highly dangerous substances until industry, so slow to protect our needs, has demonstrated ways of destroying their wastes, or of making them innocuous to all segments of the ecology.

Certainly this is not to suggest that we demolish the chemical industry. That is a ridiculous proposal, and assuredly an unwanted one. It is after all the chemical manufacturers who provide 6 percent of our gross national product, whose labor force runs into the hundreds of thousands of workers, and whose products are essential in many aspects of our lives. Nevertheless, there is need for decreased production of many compounds. The technology could be made available, if industry were to be pressed for replacement of many of the pernicious chemicals—used in such things as pesticides, plastics, and synthetic wear—by other less malign ingredients.

There is in the end one fundamental thought: our lives are dependent upon an earth which can, fortunately, absorb considerable abuse but whose limits of resilience have been exceeded in numerous places. Our arrogance and our science have implanted in the life-sustaining soils and waters toxic substances with which they, and we, cannot contend. Only when we acknowledge our folly and temper our greed will our society begin to conform to the needs of the nature outside and inside ourselves. And only then will we be sure that what rises from the ground or what is in our air and rivers will be, as it ought, the source of life and good health—and not the agents of an untimely death.

INDEX